TWELVE SECONDS
TO THE MOON

TWELVE SECONDS TO THE MOON

A Story of the Wright Brothers

Second Edition, Revised and Enlarged

Rosamond Young and Catharine Fitzgerald

United States Air Force Museum Foundation, Inc.
Dayton

Dedication

This book is dedicated
to the
United States Air Force Museum Foundation

The purpose of the Foundation is to preserve
the heritage and traditions of the
United States Air Force

Board of Managers, 1983

William S. Anderson
James F. Barnhart
John W. Berry, Sr.
E. Bartlett Brooks
The Honorable Clarence J. Brown
Dr. Robert T. Conley
The Honorable Tony Hall
Paul L. Hyde
Richard J. Jacob
Jervis S. Janney
Virginia W. Kettering
Dr. Kalman L. Levitan
Robert S. Margolis
General Jack G. Merrell USAF (Ret.)
Dr. Lionel H. Newsom
General James P. Mullins
Robert S. Oelman
Ervin H. Nutter
Louis Polk
C. Frank Scarborough
John F. Torley
Dr. Peter D. Williamson
John D. Woods
Louis Wozar
Dr. Richard H. Kohn
Brigadier General Richard F. Abel
Colonel Richard L. Uppstrom
Richard E. Baughman

Lieutenant Colonel Frederick C. Wolf USAF (Ret.),
Executive Secretary, Air Force Museum Foundation

Acknowledgments

Twelve Seconds to the Moon was first published in 1978 by *The Journal Herald,* Dayton, Ohio. The authors of the revised edition are grateful to Arnold Rosenfeld, editor of *The Journal Herald* and the *Dayton Daily News,* to Dennis Shere, president, *Dayton Newspapers, Inc.,* and to William Worth, a former managing editor of *The Journal Herald,* for assigning the copyright and giving permission to use photographs and plates from the original book.

Thanks are due to a number of persons for their help in the preparation of this book:

To Mr. and Mrs. Harold Miller for their reminiscences about the Wright brothers, uncles of Mrs. Miller, and to Mr. and Mrs. Horace Wright for their recollections of Horace's uncles, Wilbur and Orville,

To Dr. Patrick B. Nolan, Head of Archives and Special Collections, Wright State University Library, to Tucker Malishenko of the Wright State University Library, to Marvin W. McFarland of the Library of Congress, to the staff of the Dayton and Montgomery County Public Library, to the reference library staff of Dayton Newspapers, Inc., to Ross R. Callaway of Greenfield Village, to Russell Blythe of Carillon Park, to Joseph A. Usellis, Executive Director of Educational & Musical Arts, Inc., and to Lieutenant Colonel Frederick C. Wolf USAF (Retired), Executive Secretary, Air Force Museum Foundation, Inc.

To Milton Caniff for his portraits of Orville and Wilbur Wright,

To Frank Pauer for his art work.

We are indebted to the following for photographs: The United States Air Force Museum, Dayton, Ohio; Melba Hunt of the Kettering and Moraine Historical Society; The Smithsonian Institution, the reference library of Dayton Newspapers, Inc., L. M. Berry and Company, William Koehler, and the Library of Congress.

That all these persons and institutions were so helpful speaks eloquently of their great respect for Dayton's two foremost sons.

Rosamond Young and Catharine Fitzgerald
Dayton, Ohio, 1983

Contents

1

Chartless

Columbus found a world, and had no chart...
—Santayana, *O World Thou Choosest Not*

At seven o'clock on a morning in late September 1876 the Milton Wright family of 7 Hawthorn Street, Dayton, Ohio, gathered in their dining room. Milton, forty-eight years old, a short, stout, bearded man wearing a clerical collar, stood at the head of the table and said, "Our son Orville for the past three weeks has strayed from the way he is expected to go."

Milton opened his leather-bound Bible to a page marked by a ribbon. "The text this morning is Proverbs 22:6," he said. " 'Train up a child in the way he should go, and even when he is old he will not depart from it.' " He looked at his family. At his left sat fifteen-year-old Reuchlin, fourteen-year-old Lorin, and two-year-old Katharine. At his right sat Wilbur, nine, and Orville, five. The mother, Susan Wright, in her forty-fifth year, was a plain woman, her hair combed back into a tight knot. Her strong nose and firm mouth were softened by her deep-set large eyes.

"Let us pray."

Susan and the boys pushed their chairs back and stood. Then they knelt, backs to the table, elbows on chair seats, heads bowed. Katharine put her elbows on her high chair and peeked through her hands. "O Lord," said Milton, "we thank thee for one more night passed safely in this house and for another glorious morning. We know that we are all sinners in thy sight, and we ask forgiveness for our erring son. And as our Lord has taught us to say, 'Our Father, who art in Heaven, hallowed be thy name....' "

After the *Amen*, Milton said, "From now on beginning this morning Orville will walk to and from school with Wilbur. Mother, let us have breakfast."

Reuchlin and Lorin left to walk to Central High School in downtown Dayton. Wilbur and Orville walked through the back yard and turned into the alley.

Orville kicked a stone. "Did you tell on me?"

"No, I didn't, but I knew you were playing hooky. Where have you been?"

When the young Wright children were mischievous, they were sent to the closet under the stairway in the Hawthorn Street home.

"Running the sewing machine at Ed's house. I told Mrs. Sines that our mother lets us run our machine. I oiled it when it needed it except that Ed didn't have any oil."

"What did you use?"

"Dropped water off a feather into the oil holes and fixed it."

Wilbur stopped. "Orv, if that machine starts to rust and Mother finds out about it, she'll fix you. She'll put you in the closet every day for a week instead of just one afternoon like yesterday."

Punishment for the children in the Wright home was to spend time in the closet under the stairs in the hallway. The closet had a window, shelves of books, and magazines.

The boys' father, Milton Wright, had never departed from the way his parents had trained him to go. Dan Wright, a farmer, worked in an Indiana distillery until he was converted at a United Brethren revival meeting about the time he married Catharine Reeder. He was so greatly influenced by church disapproval of alcohol

Bishop Milton Wright
1828-1917

Susan Koerner Wright
1831-1889

Lorin Wright
1862-1939

Reuchlin Wright
1861-1920

Orville Wright
1871-1948

Katharine Wright
1874-1929

Wilbur Wright
1867-1912

that he quit his distillery job and refused ever afterward to sell so much as a bushel of his corn to a distillery. Liquor, tobacco, card-playing, gambling, dancing, swearing, and working on Sundays were forbidden in the Rush County farm home into which Milton Wright was born in 1828.

When Milton was twelve, Dan and Catharine Wright moved to a farm in Fayette County, Indiana, where Milton finished his public education and attended Hartsville College. At twenty-five he was still at home working on his father's farm when he experienced what was spoken of as a "call." He believed that he heard the voice of God asking him to become a minister just as Jesus called his disciples from their daily work. He joined the United Brethren Church and when he was twenty-seven, he received a license to preach in the White River Conference. Two years later—in 1857—he was ordained.

Milton Wright's Diary

May 27, 1857

I came to my father's by way of Andersonville Post-Office, where I got a letter from D. J. Flickinger, informing me that the Mission Board had appointed me to Oregon. I gave my father the letter and walked out. I lodged at my brother Harvey's.

May 28, 1857

I came home and remained all day. My mother was much affected by my going to Oregon, but was resigned. She said she had prayed that her sons might be ministers and she ought not to complain.

When Milton was preaching at Hartsville, he had met a student, Susan Catherine Koerner. She was the daughter of John G. Koerner, a maker of carriages and farm wagons, who immigrated from Germany to Baltimore in 1818. He married American-born Catherine Fry. Their daughter Susan Catherine was born April 30, 1831, at Hillsboro, Loudoun County, Virginia. The family moved to Union County, Indiana, where John Koerner constructed a twelve-building carriage factory on his farm. Susan, who learned carpentry from her father, excelled in mathematics, Greek, and Latin at Hartsville College.

Milton Wright's Diary

June 19, 1857

I went to Hartsville and dined at Professor L. Kreck's. I supped at David's, went to Daniel's and had my first talk with Susan Koerner. I asked her to go to Oregon with me.

When Milton left for the long trip to Oregon, he traveled alone. By the time he arrived in August 1857, he was so weak from malaria he contracted in Panama that he gave up his missionary appointment. He took a teaching position at Sublimity College, a church school fifteen miles southeast of Salem, Oregon.

Milton's teaching term ended in October 1859 and he returned to his parents' home in Indiana on November 14.

Milton Wright's Diary

November 15, 1859

I visit Harvey's in the morning. I go in the afternoon to John G. Koerner's to see Susan. It is rainy and it is night before I get there. Daniels' are at supper and Silas Koerner and wife are at the table with them. I sat down by the fire. At last Elmira came and said, "I believe it is Brother Wright." Daniel went to tell it at the other house. Rev. Franklin Morgan was there and came back with Daniel and expressed great joy at meeting me. I go to the other house and talk with Susan till a late hour.

Milton and Susan were married at Koerner's farm Thanksgiving Day, November 24, 1859. From the day of their wedding until the end of the year Susan accompanied Milton on his itinerant preaching trips in Indiana. They stayed in many homes—at the Koerners, at the Wrights, at Milton's brothers, and at homes of church members. Milton chopped wood, built a sleigh, and took Susan to Connersville to shop for household goods.

Milton Wright's Diary

December 28, 1859

We load up furniture, take Daniel's wagon and team and I reach my father's late. It was so icy that the team could hardly climb the hills.

January 1, 1860

Susan and I go to Bloomings Grove and dine at a hotel, had a goose forty years old. Lodged at Father's. It is awfully cold. We warmed several times. The goose was as tough as a crow.

January 2, 1860

Susan and I went early to our home in New Salem, Rush County, Indiana, and I began school at eight-thirty. It had three rooms, and it was a very happy home.

For the next eight years Milton preached and taught in several Indiana counties. Susan and Milton's son Reuchlin was born March 17, 1861, in Grant County; Lorin November 18, 1862, in Dublin County; and Wilbur April 16, 1867, in Henry County.

Milton jotted in a notebook "Willy had been walking since he was nine and a half months old. Lorin was about eleven months old and Reuchlin was twelve months old when he commenced walking."

At the general conference of the church in 1869 Milton was appointed editor of *The Religious Telescope*, a church paper published in Dayton, Ohio.

The United Brethren Church had a disagreement over membership in the Freemasons, the radicals believing that membership in the Masons was reason for expulsion from the church while the liberals—although they believed Freemasonry was unchristian—thought that Masonic membership should not be grounds for expulsion.

Milton, a radical, accepted the editorship of the *Telescope*, gave up farming and teaching and moved to a rented brick house at the corner of Sprague and Third Streets in Dayton.

A year later he bought the home at 7 Hawthorn Street, where Orville was born August 19, 1871, and Katharine on Orville's birthday three years later. Susan taught the boys to cook, sew, and use hand tools. Milton provided them with books and encouragement to satisfy their curiosity. Both parents required the boys to earn their spending money. Wilbur wiped the supper dishes for a penny, and Orville collected bones and old rags and sold them to a ragman.

Milton Wright had ambitions to be elected to the highest post in his church—that of bishop. To that end he entertained in his home all clergymen who came to the publishing house. He wrote their names and details about their families in a notebook so that when he met them again, he could ask about the wives and children by name. Whenever he traveled on church business, he took lunch at one home, dinner at another and stayed overnight at a third.

At the quadrennial convention of 1877 his efforts succeeded. He came home to Dayton with a four-year appointment as bishop of the Mississippi Valley Conference.

The position required Milton to travel six months of the year and to move the family to Cedar Rapids, Iowa. "We won't sell our house, though," he told the family. "If I am not reelected four years from now, we will have this house to come back to. Meanwhile, we'll rent it and hope that the rent will be enough to pay for a house in Cedar Rapids."

Orville contracted diphtheria in November 1878. Milton wrote in his diary for November 7, "Orville is still sick, but seems not specially dangerous till in the night. Then he has sinking spells and seems to be nearly gone." Before Orville recovered, Lorin also contracted diphtheria and then Reuchlin came home ill from the same disease. Milton left on a preaching trip for Moline, and Susan looked after three sick boys and young Katharine. Milton did not return for Christmas, but three days later dragged into the house, feverish and weak with diphtheria.

When Lorin joined Reuchlin at Western Academy in the fall of 1879, Wilbur, who was twelve, turned for companionship to eight-year-old Orville. The two were playing marbles one afternoon when their father returned from a trip to Ohio. He set down his valise and took out an object made of paper, bamboo sticks, and cork. "Here's a present for you, boys." He turned a stick that twisted a rubber band, fastened it, and tossed the object into the air. Both boys reached out to catch it. Instead of falling, it fluttered and flew several feet across the room.

When it settled to the floor, Wilbur picked it up. "What is it?"

"It's called a helicopter, from the Greek *helix*, meaning *stem-like*, and *-pter*, which means *wings*." The rubber band furnished the motive power. As the band untwisted, the helicopter flew several seconds. The boys played with it until it fell to pieces.

"I'll make a bigger one and then it will fly farther," said Wilbur. Although he built several models, none of them would fly.

When Orville was seven, he organized a dozen schoolmates into an army, appointing himself general. One day he coaxed Wilbur into defending a woodshed behind their house. The army attacked with bullets made of putty. When there was no return fire, the boys approached the shed door and stood in a group while Orville gave instructions how to storm the fortress and take the lone defender prisoner. While he talked, the general fell to the ground. The troops bolted. Wilbur had run a clothes prop under the door and flattened his brother.

Orville's class was dismissed early one day. Orville led the army around the building, and ordered his men to throw gravel at the windows of the classrooms still in session. The janitor saw them and gave chase, but the army escaped. Orville warned his troops that if the janitor told on anybody, he would probably have to stay after school. "If that happens," he said, "we will all stay. One for all and all for one."

The parlor at 7 Hawthorn Street—photo taken at Greenfield Village, Dearborn, Michigan

At the close of school next day the teacher said, "Orville, please remain after class." When the class left, Orville's army stayed in their seats. "Come to my desk, Orville," the teacher said. When Orville rose, the army rose with him. They marched in a body to the desk. "What are you boys doing here? I asked only Orville to stay."

"One for all and all for one," said Orville.

"I don't understand, but what I want to know, Orville, is whether you have your piece ready to say next Friday for the parents."

The day Orville passed from the Second Reader to the Third, the bishop took the boys to have their pictures made. The face of thirteen-year-old Wilbur had a lean, smooth look. His dark hair had already begun to thin along the part. Orville's face had not yet at eight lost some of its childlike roundness. Reuchlin, nineteen, had passed his teacher's exam and left for a teaching position not far from Cedar Rapids. Lorin, eighteen, was in his second year at Western College.

Susan, forty-nine years old, had not been well for several months. She tired easily and some days when the boys came home from school, they found their mother lying down. At those times Wilbur cooked supper. He fried sausage, ham or round steak, made mashed potatoes and brown gravy.

Whenever Milton had to be away, he kept the family informed of his whereabouts by post cards and letters. Generally the letters were addressed to Susan and contained special messages for the boys and Katharine, but one day a card came addressed to Mr. Orville Wright. No sooner had Orville read it than he asked for a card and pen. He spent a quarter of an hour at the parlor desk and then showed what he had written to his mother.

Dear Father

I got your letter today. My teacher said I was a good boy today. We have 45 in our room. The other day I took a machine can and filled it with water then I put it on the stove and I waited a little while and the water came squirting out of the top about a foot. The water in the river was up to the cracker factory about a half a foot. There is a good deal of water on the island. The old cat is dead.

Your son

Orville

When Milton returned home, he brought Orville's card, dated April 1881, and put it in a drawer along with his sermon notes, diaries, and articles for *The Religious Telescope.* It is the first entry in the Wright Collection in the Library of Congress.

At the general church conference in 1881 the liberal faction gained control of the delegates and consequently, Milton lost the election for bishop.

The Wrights moved to Richmond, Indiana, in June. The move brought them nearer Milton's farm. Susan's sister and brother-in-law, Caroline and Daniel Zeller, lived in Richmond.

Susan's health concerned her family. The doctor diagnosed her illness as consumption. Since they lived near the Zellers, Emma Zeller and sometimes Caroline helped Susan with the housework. This eased Milton's mind somewhat, particularly as presiding elder, he had to be away from home months at a time.

Milton Wright's Diary

March 14, 1882

Susan took a hemorrhage (perhaps bronchial) in the evening. Went and got medicine of Dr. Hibberd.

March 17, 1882

Susan gaining a little. At home all day. Did not go to my quarterly meeting on account of Susan's sickness.

In Richmond Orville made and flew better kites than those sold in the stores, and he earned pocket money by selling them to his friends. He also collected scrap iron and steel and sold it; in this enterprise he enlisted the help of seven-year-old Katharine. When he found a piece of metal he wanted, he made her ask the owner for it because he was too bashful.

One venture that failed was making chewing gum, which he intended to cut into little cakes, wrap in paper and sell to his friends. He mixed tar with sugar and flavoring from the kitchen and cooked the mixture in a black iron kettle over a fire in the back yard. He sampled the product while it was cooking, became ill, and abandoned the project.

Wilbur and Orville built the porch and made the railings and posts for the home at 7 Hawthorn Street, Dayton.

While Orville busied himself with money-making schemes, Wilbur read books in his father's library. Whenever his mother needed him to help with the cooking or laundry, she could always find him near the bookshelves.

In 1882 Milton started *The Richmond Star,* a monthly religious publication. The special purpose of the *Star* was to give information and argument against Freemasonry, in opposition to *The Religious Telescope,* the official U. B. church publication.

Both Wilbur and Orville helped their father by folding copies of the *Star* as they came from the press, packaging, and taking them to the post office.

Two years after the move to Richmond, Reuchlin and Lorin returned to Dayton to work. In June of 1883 the rest of the Wrights moved back to Dayton, renting a house at 114 North Summit Street until their own home on Hawthorn Street could be vacated and the walls be repapered. Milton continued to edit the *Star* in Dayton, making frequent trips to Richmond to read proofs, make up the pages, and mail the paper.

Neither Wilbur nor Orville finished the school year in Richmond. Wilbur had completed his senior year in high school, but he did not return for his diploma. Orville had no report card to show he had completed the sixth grade, but in September he entered the seventh grade at Garfield School. Wilbur enrolled at Central High School for a postgraduate course in Greek and trigonometry.

He belonged to the high school ice hockey team. In a game with sons of staff officers at the Soldiers' Home, a player lost control of his stick and smashed Wilbur in the face, knocking out four upper front teeth. A surgeon at the Home gave him first aid. Wilbur refused his offer of transportation home saying that if someone brought him, his mother would be frightened.

The injury had severe effects. While he was having dental repair, his diet was restricted to liquids, toast and eggs. He became so sensitive about his appearance that he gave up plans for college and dropped his friendship with a girl at Central High. He stayed at home, reading in the family library and helping his ailing mother.

Although the liberal element of the church grew more powerful every year, yet at the quadrennial conference in 1885, Milton was again elected bishop and was assigned to the Pacific Coast District. He commented, "They want to get this old radical as far out of the way as possible."

Susan asked, "Do we have to move again?"

Milton shook his head. "Much as I would like to have you and the children with me, it is unfair to uproot you again. No, we shall keep this house as our home and I will be the one to do the traveling." Before he left for the coast in May, he hired a housekeeper to help Susan. He did not return until January 4, 1886.

Lorin, living in Kansas, wrote to Katharine in 1888, "What does Will do? He ought to do something. Is he still cook and chambermaid?" Wilbur was. He had made a rocking chair for his mother, bedridden since 1887, and whenever she was strong enough he carried her downstairs to sit for a while. After supper he carried her back to her bedroom.

Milton Wright's Diary

October 25, 1888

Wilbur is out the third time circulating, in two wards, notices to Republicans to register for Harrison's election.

Christmas was a high point of the year for the Wrights. Milton came home for Christmas in 1888 and so did Lorin, who had been away for eighteen months. The family decorated the tree with popcorn strings, spun glass ornaments from Germany, small woven paper baskets, and candles set in fluted tin holders clamped to the balsam branches. Under the tree lay one present for each family member.

Milton Wright's Diary

December 25, 1888

Reuchlins' come over in the morning to get Katharine's presents — two cloth dogs and a dress. Picture frame for them, etc. Reuchlin a gold pencil; Lorin's present, a white silk muffler; Wilbur's, an alpaca umbrella; Orville's, $3.50; Katie's, new dress and handkerchiefs; Mother's, red knit shawl and glass tumblers.

The family took dinner at Reuchlin's house. Wilbur stuffed the turkey, baked it, and carved it, saying, "Ah, 'tis a fine beastie."

Orville used his $3.50 Christmas present to buy type. Four years before, he and his neighborhood friend Ed Sines had started a job printing business with Ed's toy printing press. They printed business cards because the press printed only one line of type at a time. When in 1881 the bishop saw the small press, he suggested to Lorin and Wilbur that they should trade their home-made rowboat for a larger press for the boys. They made the trade and gave Orville and Ed a three-by-five-inch self-inking press. The bishop gave them thirty pounds of type. The gifts enabled the boys to do more job-printing and also start a newspaper for their high school class.

In one of their early issues they ran out of news. They filled an entire page with "Sines & Wright Job Printing." When the bishop saw it, he suppressed the paper. Said he, "People will think you are lazy."

One customer paid a printing bill in popcorn. Ed wanted to eat the payment.

Orville argued that they should sell the popcorn and put the money back into the business. Neither boy would give in to the other. As a result, Orville bought Ed's interest for one dollar and then hired him to work for the company.

When Orville was seventeen in 1888, he tried to build a still larger press with scrap iron and wood from the backyard lumber pile. He ran into problems he was unable to solve, and Wilbur suggested solutions that while unorthodox worked well. For one thing, he proposed using the folding mechanism from an old buggy top to operate a gravel-filled roller to press the type against the printing surface.

As business increased, Wilbur became interested in Orville's project. The two built a still larger press, one that produced an eleven-by-sixteen inch page. Orville withdrew from high school and rented a room at 1210 West Third Street, a five-minute walk from home. On March 1, 1889, they began publication of a weekly, *The West Side News.* At first the subscription price for the three-column paper was forty cents a year or ten weeks for ten cents. By April when it became a four-column paper, the price increased to eighty cents a year or six weeks for a dime.

The publication carried news of foreign affairs, national news, local items, and advertisements. Wilbur wrote editorials; Ed Sines sold advertisements.

"This will be the first and last number of the *News* within President Cleveland's term," Orville wrote in the first issue. After a description of Mrs. Benjamin Harrison's inaugural gown, he concluded, "Mrs. Harrison received about 25 letters a day begging her to intercede with her husband to secure offices for the writers. All such letters speedily found their way into the wastebasket."

At times Wilbur wrote sketches for the paper. "A scientific journal explains in a long article how thunderstorms come up. They wait until the Sunday School picnic reaches the grove and gets fairly to business at Copenhagen, swinging flirtation, croquet and other innocent games, and then they come up like thunder and lightning. It takes the average thunderstorm not more than ten minutes to come up in the neighborhood of a picnic."

While the brothers were occupied with their newspaper, the bishop had trouble with the church. Long in disagreement about Masonic membership, the church established a commission to revise the church constitution. The liberal faction, in favor of Masonry, demanded the drafting of a new constitution; the radical group, of which Bishop Wright was a leader, declared that no changes whatever should be made in church doctrine or method. At a conference in York, Pennsylvania, in May 1889 the churchwide vote upheld the liberals. Milton and fourteen other radicals walked out of the conference and in another building declared themselves the true governing body and proceeded to conduct the business of the church. This brought on a split in the denomination. Of about two hundred thousand members, some sixteen thousand followed Bishop Wright.

Milton Wright's Diary

July 3, 1889

Rev. Wm Miller called and wanted me to go to Bellefontaine to see Judge Lawrence for him. Susan said I ought to go, and I went. Returned at 6:00. Called at Rev. Wm Dillon's and at Dr. L. Davis.' Susan slept well tonight. Awake at about 1:00 and got Susan a cool drink.

July 4, 1889

About 4:00 I found Susan sinking and about five awakened the family. She revived about 7:00 somewhat, but afterward continued to sink till 12:20 afternoon. Mary Caroline Zeller came.

July 5, 1889

Bought a beautiful lot in Woodland, $135. Made arrangements as best I could for the funeral. Wilbur had me change to a new lot.

July 6, 1889

D. K. Zeller and Emma came and Daniel Koerner, and our son Reuchlin. Funeral at 2:00 in the afternoon. Bishop Halleck Floyd preached. Bury Susan in Woodland Cemetery about 4:00 p.m. Daniel Koerner and Emma Zeller stayed with us.

Heavy black rules set off the two center columns of the July 8, 1889, editorial page of the *West Side News*. Except for the masthead—Wilbur Wright, Editor, Orville Wright, Publisher—the outside columns were blank. In the center columns under the heading "Our Mother" Wilbur wrote in part:

After many years of affliction so heroically borne, our mother has gone from among us.

Mrs. Susan Catherine Wright, wife of Bishop Milton M. Wright, died at noon on the Fourth of July at the family residence on Hawthorn Street in the fifty-ninth year of her life.

We children learned to look upon mother as almost perfection itself. No kinder mother ever lived than ours; none who loved her children more; none who more unselfishly sacrificed her own comforts and joys to give pleasure and happiness to those she loved.

For nearly eight years, she has been afflicted with lung disease and has gradually declined in health, but in that time no one ever heard one word of complaints pass her lips. Her clearness of mind, patience and endurance have been remarkable, and her courage and fortitude in these years of affliction have the more endeared her to her family.

Wilbur was twenty-two, Orville almost eighteen, and Katharine nearly fifteen when their mother died.

In April 1890 Orville and Wilbur turned *The West Side News* into a daily, *The Evening Item*. But because they were competing with established, city-wide newspapers, they found a neighborhood daily was not financially successful and dropped it in August. They produced pamphlets, advertising booklets and *The Tatler*, a journal edited by Paul Laurence Dunbar, a high school classmate of Orville. One day Dunbar chalked a ditty on the shop wall:

> *Orville Wright is out of sight*
> *In the printing business.*
> *No other mind is half so bright*
> *As his'n is.*

The 1890's has been called the decade of the bicycle. For years bicycles were difficult and dangerous to ride because the front wheel was five feet high while the back wheel was only about eighteen inches high. A British manufacturer introduced a new safety bicycle with two equal-sized wheels and pneumatic tires.

Bicycle fever stuck Americans. During the 1890's ten million of them bought the new vehicles. Everybody took to wheels—doctors, clergymen, women, sportsmen. Fifty-three-year-old Frances E. Willard, founder of the Woman's Christian Temperance Union, announced in 1892 that she would learn to ride a bicycle, partly to interest men in leaving the saloon for the more wholesome air of the high road.

Orville bought a Columbia for $150, and Wilbur bought a secondhand Eagle for $80. Orville entered a few bicycle races with Wilbur as starter and brought home a number of trophies. But soon they realized that money was to be made in

Bicycle shop at
1034 West Third Street,
Dayton, 1893

repairing bicycles and selling new ones, accessories, and parts. They rented a shop at 1005 West Third Street not far from their printing shop and stocked it with seven popular makes of bicycles.

Since the bishop was still away from home much of the time, Lorin had married and moved to nearby Horace Street, Reuchlin and his family had moved to Kansas, and Katharine had left for a year of preparatory school at Oberlin, the brothers kept house for themselves. "We have been living fine," wrote Wilbur to his sister in September 1892. "Orville cooks one week and I cook the next. Orville's week we have bread and butter and meat and gravy and coffee three times a day. My week I give him more variety. You see by the end of the week there is a lot of cold meat stored up, so the first half of my weeks we have bread and butter and hash and coffee, and the last half we have bread and butter and sweet potatoes and coffee."

The following year they moved the bicycle shop to larger quarters at 1034 West Third Street. Although they were now the owners of two businesses, they found time to remodel the home at 7 Hawthorn Street. They added a front porch with a railing and posts. They rearranged several of the rooms inside and added two gas fireplaces, one in the parlor, one in the living room. They did all the masonry, carpentry, plastering, woodturning, and painting.

"The bicycle business is fair," Wilbur wrote to his father in September. "Selling new wheels is about done for this year but the repairing business is good and we are getting about $20 a month from the rent of three wheels. We have done so well renting them that we have held on to them instead of disposing of them. Could you let us have about $150 for a while? We think we would have it all ready to pay back by the time you get home."

He added that in the years after his hockey accident when his health had been poor, he had felt going to college was a waste of time. Now that his health was better, he was thinking of becoming a teacher. "I do not think I am specially fitted for success in any commercial pursuit even if I had the proper personal and business influence to assist me. I might make a living but I doubt whether I would ever do much more than this. Intellectual effort is a pleasure to me and I think I would be better fitted for reasonable success in some of the professions than in business."

The bishop sent the $150 and assured Wilbur that he would help him through college.

One day when he came home from the bicycle shop, Wilbur picked up from the hall mail table the September 1894 *McClure's Magazine.* As he leafed through it he saw photographs of a flying object that reminded him of the helicopter his father had given him and Orville sixteen years before. These were not, however, photographs of a toy, because hanging beneath the wings was a man and he was *flying.*

The captions identified the man as Otto Lilienthal, who had been studying aerial navigation at his home in Germany for twenty years. His teacher was the larger flying birds. His study had convinced him that it was the parabolic curve of a bird's wing that enable it to fly. "A bird flies with the wind," the article quoted him, "but he sails or soars against it."

After years of experimenting, Lilienthal constructed a set of wings made of collodion-coated muslin stretched over split willow frames. The wings imitated the three parts of a bird's wing. His apparatus, as he called it, consisted of the arched

Otto Lilienthal
1848—1896

wings, which folded somewhat like an umbrella, a vertical rudder shaped like a flat hand fan to keep the glider facing the wind, and a horizontal rudder to prevent a sudden change in balance.

He fastened himself into the apparatus with his weight resting upon his elbows. With the wings folded, he climbed to a plateau he had constructed, ran against the wind and when he had reached as much velocity as he could, he jumped into the air, spreading the wings at the same time. He had accomplished flights of three hundred yards.

He had attempted to power the apparatus with a one-cylinder engine that could be operated by hand or by vapor from liquid carbonic acid. But he underestimated the power of the engine, which worked the wings with such vigor that they broke. At the time the magazine story appeared, he was building another glider.

Wilbur handed the article to Orville. That evening and many evenings afterward talk about Lilienthal, flying, birds, and parabolic curves rattled the parlor air.

The bicycle business outgrew their shop, forcing them to move to a larger space at 22 South Williams Street. In the new shop they not only repaired bicycles but also manufactured their own brands and within a year had on the market three models: the Van Cleve, the St. Clair, and the Wright Special.

Fig. I. — GLIDING MACHINE.   p. 33.

Two Lilienthal gliders; Wilbur's notes in margin of Aeronautical Annual, 1897

Milton Wright's Diary

September 4, 1896

Reach Dayton at 4:30. Found Orville very sick with typhoid fever. The temperature one time, days ago, ran to 105.5 degrees. Temperature is now about 102 or 103.

September 7, 1896

Orville was better in the morning than any day since his sickness.

October 8, 1896

Orville had tapioca today for the first time. He has lived for six weeks on milk, with a little beef broth for a couple of weeks past. He also sat up in bed for the first time in six weeks.

A few days later Wilbur walked into Orville's bedroom. "I have some bad news," he said. "When it happened you were too sick for me to tell you. Otto Lilienthal was testing a new glider near Rhinow on August 9. The glider went out of control when a gust of wind hit it and he crashed. His neck was broken and he died the next day."

The brothers had thought when they read and discussed the Lilienthal article two years before that shifting the weight of the operator was not the way to maintain balance in a glider.

That evening Orville and Wilbur pulled from their father's bookcase E. J. Marey's *Animal Mechanism: A Treatise on Terrestial and Aerial Locomotion.* In it Marey discussed many apparatuses that had been devised to study the movement of a bird's wings in flight. Both Marey and Lilienthal agreed that if the flight of birds was studied in a scientific manner, hope was assured of man's being able to imitate their method of locomotion.

When they finished Marey's book, they walked to the Dayton Public Library. They found no books on the subject of aeronautics, but they took home several books on marine engineering, in the hope of finding an explanation of the theory of the screw propeller. The books, however, were of no help because marine engineers at the time did not base their calculations of screw propellers on a theory but on empirical formulas. When Orville asked the librarian why he did not have any books on aeronautics, the librarian said scientists held the idea in great discredit and it was therefore not a subject on which libraries spend money.

During the winter of 1899 the brothers read all they could find about the subject of flying, but it was not enough.

Wilbur Wright to the Smithsonian Institution

Dayton, Ohio, May 30, 1899

I have been interested in the problem of mechanical and human flight ever since as a boy I constructed a number of bats (helicopters) after the style of Cayley's and Penaud's machines. My observations since have only convinced me more firmly that human flight is possible and practicable. It is only a question of knowledge and skill just as in all aerobatic feats. Birds are the most perfectly trained gymnasts in the world and they are specially well fitted for their work, and it may be that man will never equal them, but no one who has watched a bird chasing an insect or another bird can doubt that feats are performed which require three or four times the effort required in ordinary flight. I believe that simple flight at least is possible to man and that the experiments and investigations of a large number of independent workers will result in the accumulation of information and knowledge and skill which will finally lead to accomplished flight. I am about to begin a systematic study of the subject in preparation for practical work to which I expect to devote what time I can spare from my regular business. I wish to obtain such papers as the Smithsonian Institution have published on this subject, and if possible a list of other works in print in the English language. I am an enthusiast, but not a crank in the sense that I have some pet theories as to the proper construction of a flying machine. I wish to avail myself of all that is already known and then if possible add my mite to help on the future worker who will attain final success. I do not know the terms on which you will send out your publications but if you will inform me of the cost I will remit the price.

When Wilbur showed the letter to Orville, the younger man's eyes showed displeasure.

"What's wrong?"

"You shouldn't have said 'I' all the way through. You should have said 'we.' After all, we are doing this together and I am just as important as you are."

"You are right. Want me to do it over again?"

"No, it's no use wasting paper. But after this, remember."

In response to Wilbur's letter, Richard Rathbun, assistant secretary of the Smithsonian, sent four pamphlets: *On Soaring Flight* by E. C. Huffaker; *The Problem of Flying and Practical Experiments in Soaring* by Otto Lilienthal; *Story of Experiments in Mechanical Flight* by Samuel P. Langley; and *Empire of the Air* by Louis-Pierre Mouillard. He also recommended *The Aeronautical Annual* for 1895 through 1897; *Progress in Flying Machines* by Octave Chanute; and *Experiments in Aerodynamics* by Samuel P. Langley.

Wilbur ordered the books and while they waited for them to come, they read and discussed the pamphlets. The one that inspired them most was Mouillard's because although he himself had not successfully soared in his glider, yet his writings conveyed the same enthusiasm Wilbur and Orville felt. "If there be a domineering, tyrant thought," Mouillard wrote, "it is the conception that the problem of flight may be solved by man. When once this idea invades the brain, it possesses it exclusively. It is then a haunting thought, a walking nightmare, impossible to cast off."

The list of men who had already tried to fly and failed was impressive, from Leonardo da Vinci to Sir George Cayley, who built a machine with flapping wings, to William Henson, who attached a propeller driven by a steam engine to a fixed wing.

Alexander Graham Bell had tried and failed, and even Thomas A. Edison quit early in his efforts to build a flying machine. Hiram Maxim, inventor of the machine gun, spent $200,000 on a machine with a wing structure based on kite principles and a 360-horsepower engine. The machine, which weighed 8,000 pounds, smashed the guide rails the first time he turned on the engine. The entire contraption collapsed, and Maxim gave up.

When in prehistoric days the glacier moved down through Ohio, much of the ice melted in areas south of Dayton, depositing great loads of till, a mixture of clay, gravel, and boulders. One of the largest till deposits is several miles southwest of Dayton on the north side of the great bend in the Miami River. About eighty

Samuel P. Langley
1834—1906

feet of till had been exposed to years of rain. In the 1880's several ridges rose into acute peaks. Known as the Pinnacles, the area became a favorite spot for picnicking. Wildflowers grew thick in the spring and in the sky buzzards and hawks wheeled and soared.

Here on Sunday mornings, having ridden out from town on their bicycles, the brothers lay on their backs in the long grass watching the birds soar, turn, and glide. They learned that birds do not soar in a calm. They soar facing into a wind because air streaming past the upper and lower surface of the curved wings creates a lifting force on the underside of the wings.

They observed that birds twist their wing tips, changing the angle of the leading edge presented to the wind. What they had to do, the brothers concluded, was to find a way to change the angle of a glider wing just as a buzzard changes its angle of wing.

The first method that occurred to them was to pivot glider wings on the right and left sides on shafts carrying gears at the center. The meshed gears would cause one wing to turn upward in front when the other wing was turned downward. They thought this method would give a greater lift on one side than on the other so that it would not be necessary to shift the operator's weight to maintain balance. They could not, however, envision any method of building such a device which would be strong enough and at the same time light enough to be practicable.

One day when Orville, Katharine, and Harriet Silliman, Katharine's college classmate at Oberlin, were away from home, a customer came into the bicycle shop to buy an inner tube. Wilbur took the tube from its long narrow cardboard box and stood twisting the box while he and the customer talked.

Wilbur looked down at the box. As he saw his hands twisting it, in his mind the box became a biplane with the top wing connected to the bottom wing by the sides of the box. He cut off the ends of the box so that he had a rectangular hollow box two inches by ten.

By holding the top forward corner and the rear lower corner of one end of the box between his thumb and forefinger and the rear upper corner and the lower forward corner of the other end of the box with his other thumb and forefinger and by pressing the corners together, he gave the upper and lower surfaces of the box a helicoidal twist, presenting the top and bottom surfaces of the box at different angles at the right and left sides.

"We can twist the wings of a double-deck type machine in the same manner," said Wilbur when he showed the box to Orville. "When it flies, the wings on the right and left sides can be warped so as to present their surfaces to the air at different angles and secure unequal lifts on the two sides."

Within a few days the brothers began construction of a model kite using the principle demonstrated by the inner tube box.

Wilbur first drew plans for a biplane kite with wing surfaces five feet long and thirteen inches wide. He fastened two lines to the forward corners of the upper and lower right wing tips, attaching the ends of the lines to the four ends of a pair of movable crossed sticks to be held in the operator's hands. The bottom wing lines he attached to the top of the sticks and the top lines to the bottom. By moving the sticks in opposite directions, the operator could twist the wings, he theorized, so that when one side of the kite sank, a twist of the sticks would give the wings more lift and restore the balance.

Wing warping was the Wright brothers' first great invention. It controlled lateral balance or roll of the aeroplane by enabling the operator to turn the machine to the left or to the right. To turn to the right the operator raised the left wing and lowered the right. He reversed the movement to turn left.

In a modern airplane lateral control is accomplished by ailerons on the trailing edge of the wings.

They built the kite in the back room of the shop and had it ready for testing in August 1899. "I'll stay in the shop," said Orville. "We can't afford to close it and after all, you drew the plans."

Wilbur took the kite on his bicycle out West Third Street until he came to an open field. The kite responded promptly to the warping of the surfaces, always lifting the wing that had the larger angle. Several times when he shifted the upper surface by manipulating the sticks, the nose of the kite turned downward as he had planned, but the turning created a slack in the cords. When that happened, the kite made a sudden dash toward earth. Three schoolboys, who were watching, dived to the ground to keep from being hit. "I don't understand why the kite took those dives," Wilbur said when he reported later to Orville at the shop. "It responded well to lateral control. As long as the string was tight, it flew beautifully."

Satisfied that their method would work, they decided to build a full-size glider capable of carrying a man.

Lilienthal's 1889 tables gave air pressure valves at varying angles of incidence. From the tables they calculated that a machine with a little more than 150 square feet of area would support a man in a 16-mile-per-hour wind. They planned to fly the machine first as a kite in order to gain practice with safety and a minimum of effort.

Percy S. Pilcher, an English aeronaut who had made several hundred successful glides, was killed October 1, 1899, while making a demonstration glide in England. This news, coupled with the earlier of Lilienthal's death, confirmed their judgment that shifting body weight was not the answer to maintaining equilibrium in flight.

Wilbur wrote in late December 1899 to the Bureau of Statistics, United States Weather Bureau, for a report of wind velocities for the months of August and November in Chicago or its vicinity, explaining "We have been doing some experimenting with kites with a view to constructing one capable of sustaining a man. We expect to carry the experiments further next year."

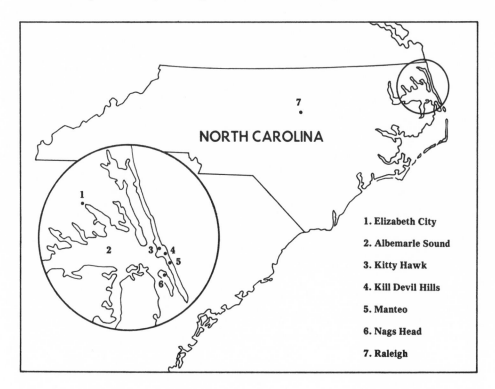

NORTH CAROLINA

1. Elizabeth City

2. Albemarle Sound

3. Kitty Hawk

4. Kill Devil Hills

5. Manteo

6. Nags Head

7. Raleigh

The brothers carried out their experiments, not near Chicago but on the east coast of North Carolina.

*At Kitty Hawk the
brothers flew their 1900
glider first as a kite, later
as a manned glider.*

2

Kitty Hawk, O Kitty

Kitty Hawk, O Kitty
Into the unknown
Into the sublime.
Off these sands of time.
　　　　　　　　—Frost, *Kitty Hawk*

Wilbur was the methodical brother. Every day at noon he stepped out the door of the bicycle shop, headed through the alley off Fourth Street, walked up the path through the back yard of 7 Hawthorn to the porch, opened the back door, took off his hat and hung it on a peg, reached up to the top of the cupboard where he kept a comb, ran it through his thinning hair, replaced the comb, washed his hands, and took a cracker from the cracker jar. That was the signal for the food to be put on the table.

Orville was the mischievous brother. He loved to tease Carrie Kayler, the fourteen-year-old girl Katharine Wright had hired to do the cooking and housekeeping. Sometimes the teasing went too far. If Wilbur saw that Carrie was about to cry, he said, "Now Orv, that's enough." When he heard that, Orville stopped. Wilbur was the only one in the house who could make his younger brother behave.

When Carrie first started working for the Wrights, she did not know much about cooking. One noon Wilbur said, "I cannot eat this any longer. I like the kind Mother always made."

That evening while Carrie was preparing supper, he went into the kitchen, took off his coat, and rolled up his sleeves. "Carrie, I want to show you how to make good brown gravy. I think you'll like it better than the white." He heated the fat in the skillet until it sizzled, mixed in the dry flour and when it was brown and smooth, poured in a mixture of water and milk, slowly stirring all the while. When it thickened, he salted and peppered it and said, "There. That's good gravy. Try a taste."

From that day on, Carrie made the kind of gravy the Wrights liked.

Of the books the brothers read on flying one that interested them greatly was *Progress in Flying Machines* by Octave Chanute. The book, published in 1894, was a collection of a series first published in *The Railroad and Engineering Journal.* It gave a history of all earlier attempts to fly and an explanation of Chanute's theories of flying. He hoped that his research, he said, would serve as a guide for other experimenters.

Chanute, who was born in Paris in 1832, had come to the United States in 1838. He became a railroad division engineer at twenty-one and worked for the Hudson River Railroad and later various railroads in the West. In 1867-1868 he planned and supervised the construction of the first bridge across the Missouri River at Kansas City. From 1873 to 1883 he was consulting engineer for the building of iron railroad bridges.

When he made a trip to England in 1875, he became interested in aeronautics after meeting Francis H. Wenham, who believed that man's flight was possible and would be achieved by scientific study.

Wilbur Wright to Octave Chanute

May 13, 1900

For some years I have been afflicted with the belief that flight is possible to man. My disease has increased in severity and I feel that it will soon cost me an increased amount of money if not my life. I have been trying to arrange my affairs in such a way that I can devote my entire time for a few months to experiment in this field.

My general ideas of the subject are similar to those held by most practical experimenters, to wit: that what is chiefly needed is skill rather than machinery. The flight of the buzzard and similar sailers is a convincing demonstration of the value of skill, and the partial needlessness of motors. It is possible to fly without motors, but not without knowledge & skill. This I conceive to be fortunate, for man, by reason of his greater intellect, can more reasonably hope to equal birds in knowledge, than to equal nature in the perfection of her machinery.

Assuming that Lilienthal was correct in his ideas of the principles in which man should proceed, I conceive Lilienthal's apparatus to be inadequate not only from the fact that he failed, but my first observations of the flight of birds convince me that birds use more positive and energetic methods of regaining equilibrium than that of shifting center of gravity.

I make no secret of my plans for the reason that I believe no financial profit will accrue to the inventor of the first flying machine, and that only those who are willing to give as well as to receive suggestions can hope to link their names with the honor of its discovery. The problem is too great for one man alone and unaided to solve the secret.

Chanute's reply came within a few days. He told Wilbur he was in sympathy with his plans. He sent several folders and suggested that if Wilbur came to Chicago, he would be glad to talk with him. If he could not come, Chanute said that he would be happy to correspond with him further.

The summer months were busy times in the bicycle business. But the brothers kept on with their reading and after-supper discussions about flying. They customarily sat beside the living room fireplace. Orville sat up straight in his chair, arms folded. Wilbur sprawled down in his chair, his hands laced together behind his head, elbows spread like wings.

When they built their model glider in 1899, they used pine for the framework and covered the wings with muslin brushed with shellac. When they started construction of their man-sized glider in August 1900, Chanute suggested they use spruce framework and instead of commercial shellac, they mix their varnish from his own recipe. Chanute's varnish caused the desirable effect of shrinking the wing coverings.

He further suggested that since they planned to experiment with their glider late in the year, they might find good locations with sand hills on the Atlantic coast in South Carolina or Georgia.

The brothers studied maps, consulted the Weather Bureau at Washington, and wrote to Joseph J. Dosher at the weather station at Kitty Hawk, North Carolina.

Dosher replied on August 16 that Kitty Hawk on the Outer Banks of North Carolina was near the top of a 175-mile stretch of coastal sand dunes from the borders of Virginia to Cape Hatteras. He said the beach was about one mile wide and clear of trees and high hills with a northeast and a north wind in September and October. There were no houses to rent; Wilbur would have to bring a tent.

Two days later a letter arrived from William J. Tate, to whom Dosher had shown Wilbur's letter. Tate described a stretch of sand one mile by five with an eighty-foot hill in the center and no bushes or trees to break the wind current. "You can reach here from Elizabeth City, N. C., direct from Manteo 12 miles from here by mail boat every Mon., Wed., and Friday," he wrote, adding that board was available if there were not too many in the party and that he would take pleasure in doing all he could for Wilbur's convenience, success, and pleasure.

When they received Tate's letter, the brothers decided on Kitty Hawk.

Early in September Wilbur wrote to his father, who was attending a conference in Indiana, that he planned to leave within a few days for North Carolina. He said, "It is my belief that flight is possible and, while I am taking up the investigation for pleasure rather than profit, I think there is a slight possibility of achieving fame and fortune from it."

Orville was to stay behind in the bicycle shop until Wilbur had the camp ready. Katharine, her brown hair drawn back into a loose knot, rimless pince nez clamped to her nose, and wearing a white shirtwaist and dark skirt, helped him pack, sending along enough clean starched collars for the stay and slipping a glass of homemade jelly into his bag. To the bishop she wrote, "We are in an uproar getting Will off. The trip will do him good. I don't think he will be reckless. If they can arrange it, Orv will go down as soon as Will gets the machine ready." Wilbur left Dayton for Elizabeth City September 6, 1900.

Wilbur Wright Memorandum

September 13, 1900

Arrived at Old Point about six o'clock p.m. the next day, and went over to Norfolk via the steamer *Pennsylvania*. Spent Saturday morning trying to find some spruce for spars of machine, but was unsuccessful. Finally I bought some white pine and had it sawed up at J. E. Etheridge Co. Mill. Cumpston Goffignon, the foreman, very accommodating. The weather was near 100 Fahr. and I nearly collapsed. At 4:30 left for Elizabeth City and put up at the Arlington where I spend several days waiting for a boat to Kitty Hawk. No one seemed to know anything about the place or how to get there. At last on Tuesday (September 1) left. I engaged passage with Israel Perry on his flat-bottom schooner fishing boat. As it was anchored about three miles down the river we started in his skiff which was loaded almost to the gunwale with three men, my heavy trunk and lumber. The boat leaked very badly and frequently dipped water, but by constant bailing we managed to reach the schooner in safety. When I mounted the deck of the larger boat I discovered at a glance it was in worse condition if possible than the skiff. The sails were rotten, the ropes badly worn and the rudderpost half rotted off, and the cabin so dirty and vermin-infested that I kept out of it from first to last. The water was much rougher than the light wind would have led us to expect, and Israel spoke of it several times and seemed a little uneasy. The boat was quite unfitted for sailing against a head wind owing to the large size of the cabin, the lack of a load, and its flat bottom. The waves which were now running quite high struck the boat from below with a heavy shock and threw it back about as fast as it went forward. The leeway was greater than the headway. The strain of rolling and pitching sprung a leak and this, together with what water came over the bow at times, made it necessary to bail frequently. In a severe gust the foresail was blown loose from the boom and fluttered to leeward with a terrible roar. The boy and I finally succeeded in taking it in though it was rather dangerous work in the dark with the boat rolling so rapidly. The waves were very high on the bar and broke over the stern very badly. Israel had been so long a stranger to the touch of water upon his skin that it affected him very much.

Because of the gale the schooner put up for the night and did not sail again until the next afternoon. At 9 p.m. Perry tied up the boat at the Kitty Hawk wharf, but by that time it was dark. Wilbur stayed on board another night.

In the morning he went ashore and inquired for William Tate, who was postmaster, commissioner of Currituck County, and a fisherman. Tate welcomed him to his home and when Mrs. Tate found that for two days Wilbur had eaten nothing but Katharine's jelly, she fried ham and eggs. "His house is a two-story frame with unplaned siding," Wilbur wrote to his father, "not painted, no plaster on the walls, which are ceiled with unvarnished pine. He has no carpets, very little furniture, no books or pictures. There may be one or two better houses here but his is much above average. You will see that there is little wealth and no luxurious living. A few men have saved up a thousand dollars but this is a savings of a long life. Their yearly income is small. I suppose few of them see two hundred dollars a year. They are friendly and neighborly and I think there is rarely any suffering among them. The ground here is very fine sand with no admixture of loam that the eye can detect, yet they attempt to raise beans, corn, turnips, etc. on it. Their success is not great but it is a wonder that they can raise anything at all.

"I have my machine nearly finished. It is not to have a motor and is not expected to fly in any true sense of the word. My idea is merely to experiment and practice with a view to solving the problem of equilibrium. I am watching my health very closely and expect to return home heavier and stronger than when I left. I am taking every precaution about my drinking water."

Orville left Dayton for Kitty Hawk September 26, taking cots, tea, coffee, sugar, and other supplies Wilbur had written were almost impossible to buy in Kitty Hawk. Katharine and Lorin took over the management of the bicycle shop.

When Orville arrived at Kitty Hawk September 28, the brothers roomed and boarded with the Tates until they set up their tent camp October 4. Katharine sent a telegram to Orville saying she had fired the young man they had hired to run the shop. Orville wrote back that he wasn't surprised, as he had expected the young man wouldn't do. He told Katharine to remit at once whenever she ordered any supplies so that she could deduct five percent for cash.

The wind at Kitty Hawk blew stronger than the brothers expected. One day and night it blew thirty-six miles an hour and the next morning the Kitty Hawkers came around to see whether the young men had been blown away. Because of the strong wind they flew their glider like a kite, loading it with chains to give it work to do and running strings to the ground with which to work the steering.

The 1900 glider was a biplane with wing surfaces 17 feet by 5 feet. The net wing surface was 165 square feet. With the operator the weight was 190 pounds. The wings were covered with bias-cut French sateen. The ribs were ash and the main crosspieces white pine. Flexible hinges joined the uprights to the wings. The glider was trussed laterally and had a front rudder but no rear rudder, no horizontal or vertical tail.

Octave Chanute
1832—1910

Wilbur and Orville lived with the Tates until they set up camp October 4, 1900.

The brothers' first camp at Kitty Hawk, 1900

Orville to Katharine

October 14, 1900

In the afternoon we took the machine to the hill south of our camp, formerly known as Look Out Hill, but now as the Hill of the Wreck. (I have just stopped to eat a spoonful of condensed milk. No one down here has any regular milk. The poor cows have such a hard time scraping up a living that they don't have time for making milk. You never saw such poor, pitiful-looking creatures as the horses, hogs, and cows down here. The only things that thrive and grow fat are bedbugs, mosquitoes and wood ticks. This condensed milk comes in a can and is just like the cream of our homemade chocolate creams. It is intended to be dissolved in water, but as we cannot down it that way, we just eat it out of the can with a spoon. It makes a pretty good but expensive dessert.)

Well, after erecting a derrick from which to swing our rope with which we fly the machine, we sent it up about 20 feet, at which height we attempt to keep it by the manipulation of the strings to the rudder. The greatest difficulty is in keeping it down. It naturally wants to go higher & higher. When it begins to get too high we give it a pretty strong pull on the ducking string, to which it responds by making a terrific dart to the ground. If nothing is broken we start it up again. This is all practice in the control of the machine. When it comes down we just lay it flat on the ground and the pressure of the wind on the upper surface holds it down so tightly that you can hardly raise it again.

The forward elevator-stabilizer is the second Wright invention. It controlled the up-and-down movement or pitch—nose up or nose down—of the glider.

Today the elevators are horizontal planes at the rear of the airplane.

After an hour or so of practice in steering, we laid it down on the ground to change some of the adjustments of the rope, when within a sixteenth of second's notice, the wind caught under one corner, and quicker than thought, it landed 20 feet away.

We dragged the pieces back to camp and began to consider getting home. The next morning we had cheered some and began to think there was some hope of repairing it. The next three days were spent in repairing, holding the tent down, and hunting; mostly the last, in which occupation we have succeeded in killing two large fish hawks each measuring over five feet from tip to tip, chasing a lot of chicken hawks till we were pretty well winded, and in scaring several large bald eagles. Will saw a squirrel yesterday, but while he was crawling about over logs and through sand and brushes, trying to get a dead shot on it, it ate up several hickory nuts, licked chops, and departed, goodness knows where.

We did have a dinner of wild fowl the other day, but that was up to Tate's. He invited us up to help dispose of a wild goose which had been killed out of season by one of the neighboring farmers. The people about Kitty Hawk are all "game hogs" and pay little respect to what few game laws they have. But wild goose, whether due to its game flavor or not, tasted pretty good after a fast of several weeks in any kind of flesh except a mess or two of fish.

Kitty Hawk is a fishing village. The people make what little living they have in fishing. They ship tons & tons of fish away every year to Baltimore and other northern cities, yet like might be expected in a fishing village, the only meat they ever eat is fish flesh, and they never have any of that. You can buy fish in Dayton at any time, summer or winter, rain or shine; but you can't here. About the only way to get fish is to go and catch them yourself. It is just like in the north, where our carpenters never have their houses completed, nor the painters their houses painted; the fisherman never has any fish.

This is a great country for fishing and hunting. The fish are so thick you see dozens of them whenever you look down into the water. The woods are filled with wild game, they say; even a few "bars" are prowling about the woods not far away. At any time we look out of the tent door we can see an eagle flapping its way over head, buzzards by the dozens—till Will is 'most sick of them—soaring over the hills and bay, hen hawks making a raid on nearby chicken yards, or a fish hawk hovering over the bay looking for a poor little fish whom he may devour. Looking off the other way to the sea, we find the seagulls skimming the waves, and the little sea chickens hopping about, as on one foot, on the beach, picking up the small animals washed in by the surf.

But the sand! The sand is the greatest thing in Kitty Hawk, and soon will be the only thing. The site of our tent was formerly a fertile valley, cultivated by some ancient Kitty Hawker. Now only a few rotten limbs, the topmost branches of trees that then grew in this valley, protrude from the sand. The sea has washed and the winds blown millions and millions of loads of sand up in heaps along the coast, completely covering houses and forest. Mr. Tate is now tearing down the nearest house to our camp to save it from the sand.

A mockingbird lives in a tree that overhangs our tent, and sings to us the whole day long. He is very tame, and perches on the highest bough of the tree (which however is only about ten feet high) and calls us up every morning. I think he crows up especially early after every big storm to see whether we are still here; we often think of him in the night, when the wind is shaking the top and sides of the tent till they sound like thunder, and wonder how he is faring and whether his nest can stand the storm.

Orville sand-scrubbing his skillet at Kitty Hawk, 1900

The sunsets here are the prettiest I have ever seen. The clouds light up in all colors in the background, with deep blue clouds of various shades fringed with gold before. The moon rises in much the same style, and lights up this pile of sand almost like day. I read my watch at all hours of the night on moonless nights without the aid of any other light than that of the stars shining on the canvas of the tent.

I suspect you sometimes wonder what we eat, and how we get it. After I got down we decided to camp. There is no store in Kitty Hawk; that is, not anything you could call a store. Our pantry in its most depleted state would be a mammoth affair compared with our Kitty Hawk stores. Our camp alone exhausts the output of all the henneries within a mile.

I believe I started in to tell you what we eat. Well, some part of the time we eat hot biscuits and eggs and tomatoes; part of the time eggs, and part tomatoes. Just now we are out of gasoline and coffee. Therefore no hot drink or bread or crackers. The order sent off Tuesday has been delayed by the winds. Will is 'most starved. But he said that when we were rolling in luxuries, such as butter, bacon, corn bread and coffee. I think he will survive. It is now suppertime. I must scratch around and see what I can get together. We still have half a can of condensed milk, which amounts to six or eight teaspoonfuls.

Four nights later when Wilbur was asleep, Orville again wrote to his sister:

It is now almost bedtime. Last night at 6:30 by our time we were just crawling into bed. Our nights in Kitty Hawk are interesting and were there not so many of them, not unpleasant. A little excitement once in a while is not undesirable, but every night, especially when you are so sleepy, it becomes a little monotonous. This is "just before the battle," sister, just before the squall begins. About two or three nights a week we have to crawl up at ten or eleven o'clock to hold the tent down. When one of these 45-mile nor'easters strikes us, you can depend on it, there is little sleep in our camp for the night. Expect another tonight. We have just passed through one which took up two or three wagonloads of sand from the N. E. end of our tent and piled it up eight inches deep on the flying machine, which we had anchored about fifty feet southwest. The wind shaking the roof and sides of the tent sounds exactly like thunder. When we crawl out of the tent to fix things outside the sand fairly blinds us. It blows across the ground in clouds. We certainly can't complain of the place. We came down here for wind and sand and we have got them.

I am sitting on our chicken coop writing this letter. The coop has never had a chicken in it yet, but we hope to have two tomorrow morning. Trying to camp down here reminds me constantly of those poor Arctic explorers. We are living nearly the whole time on reduced rations. Once in a while we get a mess of fish, and if our stuff comes about the same time from Elizabeth City—which stuff consists of canned tomatoes, peaches, condensed milk, flour and bacon & butter—we have a big blowout or, as the Africans would say, "a big full." But it only lasts a day. We are expecting to have a big blowout tomorrow when we get those two chickens. We have just appointed the Kitty Hawk storekeeper our agent to buy us anything he can get hold of, in any quantities he can get, in the line of fish, eggs, wild geese or ducks. We have had biscuits, molasses, coffee, and rice today. Tomorrow morning we will have biscuits (made without either eggs or milk), coffee and rice. The economics of this place were so nicely balanced before our arrival that everybody here could live and yet nothing be wasted. Our presence brought disaster to the whole arrangement. We, having more money than the natives, have been able to buy up the whole egg product of the town and about all the canned goods in the store. I fear some of them will have to suffer as a result.

Speaking about money reminds me of a pretty good one Tom the fisherman got off a few days ago when I asked him who was the richest man in Kitty Hawk. "Doctor Cogswell," he replied. "How much has he?" I inquired. "Why, his brother owes him $15,000," and the young fisherman thought the question settled. Tom is a small chap, that can tell more big yarns than any kid of his size I ever saw. We took a picture of him as he came along the other day on his way home from the beach with a drum almost as large as he. The drum is a salt-water fish.

It is now after eight and time to be abed. A cold nor'easter is blowing tonight, and I have seen warmer places than it is in this tent. We each of us have two blankets, but almost freeze every night. The wind blows on my head, and I pull the blankets up over my head, when my feet freeze and I reverse the process. I keep this up all night and in the morning am hardly able to tell "where I'm at" in the bedclothes. From this on we are going to run the stove at night—at least from midnight on, which heats up the tent pretty well. I have been getting in from nine to ten hours of sleep every night—except the storm nights, when I'm either up running around outside the tent or in bed awake with my clothes on expecting to see the tent get up and fly away every minute.

The brothers started for home October 23, 1900, leaving the glider partially buried in the sand. Bill Tate asked whether he might use parts of it. As a result, the Tate daughters had new dresses made out of the wing sateen covers.

After their return to Dayton, Wilbur wrote a letter to Octave Chanute giving details of their experiments at Kitty Hawk in October.

Tom Tate and catch

Octave Chanute to Wilbur

Chicago, November 23, 1900

I thank you much for your letter which I have found deeply interesting and I congratulate you heartily upon your success in diminishing the resistance of the framing and demonstrating that the horizontal position of the operator is not as unsafe as I believed. I have been asked to prepare an article for *Cassier's Magazine,* and I should like your permission to allude to your experiments in such brief and guarded way as you may indicate.

Wilbur to Octave Chanute

Dayton, November 26, 1900

It is not our intention to make a close secret of our machine, but at the same time, inasmuch as we have not yet had opportunity to test the full possibilities of our methods, we wish to be the first to give them such test. We will gladly give you for your own information anything you may wish to know, but for the present would not wish any publication in detail of the methods of operation or construction of the machine.

Our machine was of the double-deck pattern and exposed an area of 177 sq. ft. complete, including a rudder of 12 sq. ft. The position of the operator was horizontal: he lying on top of the lower plane, head to the front and face down. We found the position far more comfortable than hanging by the arms; the action of the machine was much steadier; and landing was effected in the soft sand, at speeds of 20 to 30 miles per hour, without any injury to operator, or breakage of the machine. We consider the horizontal position feasible and a decided success where landing can be effected on sand or smooth grass. Our calculations indicate a head resistance at 20 mph of less than one pound for the body of the operator in the horizontal position.

The machine had neither horizontal nor vertical tail. Longitudinal balancing and steering were effected by means of a horizontal rudder in front of the planes. Lateral balancing and right and left steering were obtained by increasing the inclination of the wings at one end and decreasing their inclination at the other. The short time at our disposal for practice prevented as thorough tests of these features as we desired, but the results obtained were very favorable and experiments will be continued along the same line next year.

When in his reply to Wilbur's letter, Chanute asked if they preferred to be spoken of as "Wright Brothers" or "W. Wright and Bro.," Wilbur answered that their preference was for "Messrs. Wilbur and Orville Wright."

*"With Capt. Tate holding the right end and his son
at the other end, and with the good wind coming up
that Kill Devil Hill, the aircraft was moved
forward at a run and then launched in the air."
—Paul E. Garber.
Wilbur lies prone in 1901 glider.*

3

And Also Mosquitoes

With a hey, ho, the wind and the rain
—Shakespeare, *Twelfth Night*

"This year when we go down to Kitty Hawk," Wilbur said one morning in 1901 in the bicycle shop office, "I think we ought to have our own anemometer. Last year we stayed close to the Weather Bureau Station, but this year we will probably change our camp site and be unable to use the government anemometer."

"It ought to be a portable one," said Orville. "I wonder what kind is best and how much it costs."

Chanute advised them that for field use the best anemometer is one with light flat vanes. He himself had tested two, one made in Liverpool and a Richard, made in Paris. He believed the Richard was the better. "I will lend them, either you like," he wrote, "when you are ready to experiment."

"In general we are not disposed to use anything not our own," Wilbur replied, "but as in the present case we are already planning to spend about all we feel we ought to on our this year's experiments, it is possible we shall call on you for the loan of this instrument."

Chanute wrote that he would like to visit them in Dayton and asked when would be a suitable time.

Wilbur to Octave Chanute

May 17, 1901

During the months of March, April, May and June our time is very closely occupied, as our business requires our attention from 12 to 14 hours daily. After July 4th the tension is considerably relaxed and after September 1st we are almost free for four months. If your trip falls within the months of May or June, we could make your visit much more pleasant and satisfactory if you could arrange to stop off and spend Sunday at our home. We are entirely free all day Sunday.

*Wilbur slides in,
landing the 1901 glider.*

One hot June Saturday night Charles E. Taylor, a machinist, dropped in at the bicycle shop to, as he called it, gas. The brothers kept the shop open late on Saturday for people who could not come during the week. "Charlie," said Wilbur, "there are just two of us in the shop and we need another hand. How would you like to come work for us?"

"How much would you pay?"

"Eighteen a week."

Charlie was earning twenty-five cents an hour at the Dayton Electric Company and the Wright offer amounted to thirty cents. Besides, since he lived at Calm and Gale Streets, he could bicycle home for lunch. "I'll take the job."

Hiring Charlie meant that the brothers could leave for Kitty Hawk earlier than September. When they wrote Chanute they were going in July and would like to borrow the anemometer, Chanute brought it to Dayton himself.

Chanute arrived June 26, 1901. His coming caused a flurry in the household. Katharine decided that for dinner the dessert would be melon and bought two. "Now when you cut them," she instructed Carrie, "if one is better than the other, be sure that Mr. Chanute has a slice of the better one."

When dessert time arrived, Carrie sliced the first melon. It was almost rock-hard. The second one was good. She cut it into small pieces so that everybody could have a serving.

After dinner Katharine charged into the kitchen. "Such small portions. I was ashamed. You should have served Mr. Chanute a piece of proper size and one for the bishop and the rest of us could have been served the other melon even if we couldn't eat it."

At the time of the Dayton visit Chanute was sixty-nine, a short, chubby person, bald with a fringe of white, curly hair around his ears. He wore a mustache and short chin beard much like the bishop's. "You do have a doctor near your camp?"

"No."

Chanute frowned. "I think you are taking a big risk not to have a doctor near. I know a young man, George A. Spratt, of Coatesville, Pennsylvania, who is studying aeronautics and has had medical training. If you are willing to board him at camp, I will pay his expenses to Kitty Hawk. I also would like to send Edward C. Huffaker down to observe your flights. He is building a glider for me at the present time. I am not satisfied with the design, but if he could test it at your camp, I will send him at my expense and pay his share of the camp expenses. He could assist you while he is there. He was for three years assistant to Samuel P. Langley in his flying experiments."

Wilbur said that he and Orville would talk it over.

After Chanute returned home, Wilbur wrote that they would be glad to give what assistance they could to Huffaker and enclosed with the letter a personal invitation to him. "As for Mr. Spratt, we could not permit you to bear the expense of his trip merely to assist us; if, however, you wish to get a line of his capacity and aptitude and give him a little experience with a view to utilizing him in your own work later, we will be very glad to have him with us, as we would then feel that you were receiving at least some return for the money expended."

Chanute assured the Wrights that both men were reliable and honest, an important consideration since the brothers had no patents. Wilbur told him that they were not uneasy on that point. They themselves had benefited from the work of others, particularly Lilienthal. "We of course would not wish our ideas and methods appropriated bodily, but if our work suggests ideas to others which they can work out on a different line and reach better results than we do, we will try hard not to feel jealous or that we have been robbed in any way. On the other hand, we do not expect to appropriate the ideas of others in any unfair way, but it would be strange indeed if we should be long in the company of other investigators without receiving suggestions which we could work out in such a way as to further our work."

The 1901 glider had the same design as the 1900 glider but it was larger. The wingspan was 22 feet and the lifting power was a little more than 290 square feet. The machine weighed 98 pounds, nearly twice that of the 1900 glider, and was the largest glider that anyone had ever attempted to fly.

An article by Wilbur on the angle of incidence appeared in *The Aeronautical Journal* in London in July, 1901. "If the term 'angle of incidence' so frequently used in aeronautical discussions," he wrote, "could be defined to a single definite meaning, *viz.,* the angle at which aeroplane and wind actually meet, much error and confusion would be averted."

After detailing what he and Orville had learned from the 1900 experiments, he concluded: "Since formulation of a principle into a rule often serves to fix it more prominently in the mind, the writer ventures the following—

"Rule. The angle of incidence is fixed by area, weight, and speed alone. It varies directly as the weight, and inversely as the area and speed, though not in exact ratio."

The July *Illustrierte Aeronautische Mitteilung* included another article by Wilbur on "The Horizontal Position During Gliding Flight." He said that Lilienthal was convinced that the upright position of the operator constituted the essential factor of safety in flight and that Chanute, Pilcher, and others agreed with him. Wilbur said that he thought that while perhaps the position made landing easier, it resulted in less control of the machine in the air. "While the horizontal position requires aid in launching, the plane travels more steadily in the air, the turning motions are slower since the flyer's body becomes part of the machine, and the inertia is accordingly greater."

He concluded that the head resistance of a flying machine is reduced a good third if the operator adopts the horizontal position and that the limits of maintaining equilibrium by shifting the operator's upright body had probably been reached. While he had found that landing in the prone position in the sand at Kitty Hawk incurred no accident to operator or machine, it might not be possible to apply the position where landings were made on rocky ground. "If other methods of preserving equilibrium are used," he said, "new ways of securing the safety of the operator must also be tested."

Chanute sent them an altimeter before they left for Kitty Hawk July 7.

The worst storm any Kitty Hawker could remember, a ninety-three-mile-an-hour nor'easter, delayed their arrival in Kitty Hawk. The first night they had to sleep in a sagging hammock at the Tates.

Next morning although the rain slashed down, they selected a site for their camp at Kill Devil Hills and pitched a tent in a skin-soaking rain. The temperature was so torrid they thought they would perish from thirst. When they drove a Webbert pump, the point of it was lost in the sand. Since the nearest drinking water was a mile away, they caught rainwater that dripped from the tent. But as they had rubbed the tent with soap to discourage mildew, the water tasted terrible.

Next day they built a shed to house their glider, using locally-bought lumber and tar paper. They put up a frame building wide enough and long enough and high enough at the eaves to let their glider fit into the building in one piece. They hinged both ends of the building at the top so that when they were propped open, they made an awning.

Edward Huffaker arrived July 18 and George Spratt July 23.

Nobody had ever mentioned Kitty Hawk mosquitoes. Orville wrote home that the agonies of typhoid fever were nothing compared to the mosquito bites. The insects chewed through underwear and socks, making lumps swell like hens' eggs. To escape them, the men went to bed at five o'clock, putting their cots under the awnings and wrapping themselves in blankets except for poking their noses out. Then the wind dropped and they started sweating. When they pushed back their blankets, the mosquitoes came in squadrons.

The second night they built tents of mosquito netting over their cots and looked forward to a peaceful night. But the mosquitoes came right through the netting. Back to the blankets they went.

The third night they built fires around the camp. Spratt said he couldn't stand the smoke and he dragged his cot out into the open. Soon he dragged it back, complaining that the mosquitoes were worse than smoke. In the morning he said it was the worst night he had ever passed through.

When the brothers completed the 1901 glider July 30, 1901, their experiments with it showed that in some respects it was inferior to the 1900 glider. For one thing the control of the machine was not so good. The resistance of the framing was almost twice as large as they expected. When the machine was flying, it did not increase in speed as the 1900 machine did. This meant that they had to find a way to get higher initial velocity.

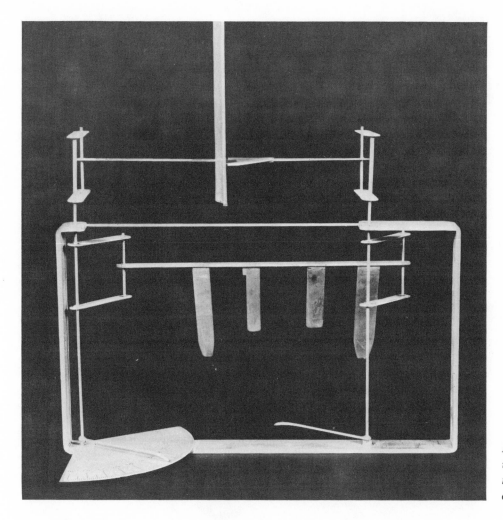

*Front view of balance
used in wind tunnel to
measure lift and drag
of different airfoils*

The machine had some good points. In forty landings it suffered no damage. After practicing only one hour they succeeded in a free flight of over 300 feet. The machine, which had over 300 square feet of surface, flew safely in winds as high as 18 miles per hour, even though previous experimenters said a machine of such size was impossible to build and manage. The lateral balance of the machine was perfect.

They spent a week remodeling the machine, reconstructing the front edge. "The travel of the center of pressure proved by experiment to travel backward instead of forward," Wilbur wrote in his diary. "Having thus proved that the travel was contrary to that desired, we trussed down the rib in the center and thus gave a shape which we hope will cause the center of pressure to move forward like a plane at all gliding angles."

After the remodeling, they made a number of good glides with a wind at five meters per second and the machine under good control.

Octave Chanute visited the camp for one week in August. In his diary notes Chanute mentioned the good control of the machine, both in straight flight and in curves. "On the occasion of the last flight made while skimming about a foot above the ground, the left wing became depressed and in shifting his body to the right to bring it up again Mr. Wilbur Wright neglected the fore-and-aft control and plunged suddenly to the ground. He was thrown forward into the rudder, breaking a number of the rudder's ribs and bruising his nose and eye."

The wind tunnel, a replica of the lost original, shows the intake with vanes to control the airflow.

The wind tunnel is the Wrights' third great invention.
Wind tunnels are used today in all aeronautical designing and testing.

Chanute asked Wilbur to give a lecture to the Western Society of Engineers about his gliding experiences. No sooner had Chanute left than a four-day rain set in. George Spratt went home. Katharine wrote to her father, "The boys walked in unexpectedly Thursday. They can only talk about how disagreeable Huffaker was."

Wilbur told his sister about the invitation to speak in Chicago. Katharine wrote to the bishop, "Will was about to refuse but I nagged him into going. He will get acquainted with some scientific men and it may do him a lot of good. We don't hear anything but flying machine and engine from morning till night. I'll be glad when school begins so I can escape."

Chanute asked Wilbur whether it would be all right to make his appearance a ladies' night. "I will already be as badly scared as it is possible for a man to be," Wilbur replied, "so that the presence of ladies will make little difference to me, provided I am not expected to appear in full dress."

A week before the lecture Katharine asked, "Will your talk be witty or scientific?"
"It will be pathetic."

"We had a picnic getting Will off to Chicago," Katharine wrote to her father a week later. "Orv offered all his clothes, so off went 'Ullam' arrayed in Orv's shirt, collars, cuffs, cuff links, and overcoat. We discovered that to some extent clothes do make the man, for you never saw Will look so swell."

He gave the speech September 18, 1901, saying in part:

The person who merely watches the flight of a bird gathers the impression that the bird has nothing to think of but the flapping of its wings. As a matter of fact this is a very small part of its mental labor. To even mention all the things the bird must constantly keep in his mind in order to fly securely through the air would take a considerable part of the evening. If I take this piece of paper and after placing it parallel with the ground, quickly let it fall, it will not settle steadily down as a staid, sensible piece of paper ought to do, but it insists on contravening every recognized rule of decorum, turning over and darting hither and thither in the most erratic manner, much after the style of an untrained horse. Yet this is the style of steed that men must learn to manage before flying can become an everyday sport. The bird has learned this art of equilibrium, and learned it so thoroughly that its skill is not apparent to our sight. We only learn to appreciate it when we try to imitate it.

He spoke of Lilienthal's being the first man to understand that balance was the most important problem in human flight. "Other men, no doubt, long before had thought of such a plan. Lilienthal not only thought but acted; and in so doing probably made the greatest contribution to the solution of the flying problem that has ever been made by one man."

After telling how he and Orville had become interested in flying, he described their experiments at Kitty Hawk in 1900 and 1901. He concluded his speech by summarizing all that they had learned about flying in gliders:

1. That the lifting power of a large machine, held stationary in a wind at a small distance from the earth, is much less than the Lilienthal table and our own laboratory experiments would lead us to expect. When the machine is moved through the air, as in gliding, the discrepancy seems much less marked.

2. That the ratio of drift to lift in well-shaped surfaces is less at angles of incidence of five degrees to 12 degrees than at an angle of three degrees.

3. That in arched surfaces the center of pressure at 90 degrees is near the center of the surface, but moves slowly forward as the angle becomes less, till a critical angle varying with the shape and depth of the curve is reached, after which it moves rapidly toward the rear till the angle of no lift is found.

4. That with similar conditions, large surfaces may be controlled with not much greater difficulty than small ones, if the control is effected by manipulation of the surfaces themselves, rather than by a movement of the body of the operator.

5. That the head resistances of the framing can be brought to a point much below that usually estimated as necessary.

6. That tails, both vertical and horizontal, may with safety be eliminated in gliding and other flying experiments.

7. That a horizontal position of the operator's body may be assumed without excessive danger, and thus the head resistance reduced to about one fifth that of the upright position.

8. That a pair of superposed, or tandem surfaces, has less lift in proportion to drift than either surface separately, even after making allowance for weight and head resistance of the connections.

Three days after the Chicago speech Wilbur wrote to George Spratt about a problem he had with Huffaker. "After you left camp, conditions which you know were none too pleasant in some respects became even worse, as they were no longer relieved by your funny stories and pleasant company, so four days after you left we also broke camp and returned home. When we came to pack up I made the unpleasant discovery that one of my blankets that had lived with me for years on terms of closest intimacy, even sharing my bed, had abandoned me for another, and had even departed without a word of warning or farewell.

"Although I regretted to part with it, yet I felt happy in the thought that its morals were safe, as it was in the company of one who made character building rather than hard labor the great aim in life! Mr. Huffaker left Sunday. He looked sheepish on departure, which I attributed to the fact that he was still wearing the same shirt he put on the week after his arrival in camp. Well, some things are rather more amusing to think over than to endure at the time.

"I enclose a few prints. That of the Huffaker machine you will please not show too promiscuously. I took it as a joke on Huffaker but afterward it struck me the joke was rather on Mr. Chanute as the whole loss was his. If ever you feel that you have not got much to show for your work and money expended, get out this picture and you will feel encouraged."

Wilbur and Orville worked out a plan to test the Lilienthal data, which they believed to be incorrect. On a bicycle they fastened a curved-wing model and a flat plate at a defined angle to each other. They mounted the wheel ahead of the handle bars on a bicycle and rode the bicycle to create a wind. They found that instead of the five degrees Lilienthal's table called for, an angle of 18 degrees was necessary for the forces to balance each other. "The advantage of the curve over the plane was so much greater than we anticipated," Wilbur told Chanute in his October 6 letter. "Although the curve was found to be far less effective than Lilienthal's table would indicate, it was so much in excess of the plane that we considered it important to obtain tests of greater exactness at smaller angles."

To make still more accurate tests, the brothers built an open-end tunnel out of wooden box 16 by 16 by 8 feet. Inside they placed a measuring device made of old hacksaw blades and bicycle-spoke wire. Through the tunnel they directed a blast of air from a fan on an old emery wheel spindle. The fan turned in a current supplied by a one-cylinder engine.

The bicycle device to test Lilienthal's data

It took them a month to get a straight wind that would vary less than one-tenth of a degree. Once they began testing, nothing could be moved in the room, and only one person could be in the room at a time. He had to stand in the same position every test time so that his movements would not create a current of air, thus invalidating the test.

In two months they tested more than two hundred small wings from three to nine inches long. They measured monoplanes, biplanes, triplanes, and even a tandem plane like Langley's. They located the center of pressure of a plane surface and measured the ratio between lift and drift. Lift is the weight-carrying ability of a wing at a given speed, and drift is its resistance to speed in motion. Their tests proved that indeed all other scientists had been wrong.

Their ingenious testing laid the foundation for all future aeronautical research.

Wilbur wrote to Chanute that they had to stop their experiments in December. "At least two-thirds of my time during the last six months has been devoted to aeronautical matters. Unless I decide to devote myself to something other than a business career, I must give closer attention to my regular work for a while."

Chanute replied that he regretted that in the interest of science, the brothers had reached a stopping place. "If, however, some rich man should give you $10,000 a year to connect his name with the process, would you do so? I happen to know Carnegie. Would you like me to write to him?

"I enclose a letter from Prof. Langley giving his reason for not publishing Lilienthal in English. As it was my understanding that the professor does not read German, I did not think that the supervision of a translator would involve much labor for him. His own book has been practically ready 3 or 4 years, and I sometimes have felt he was keeping other students back by not publishing it."

Wilbur Wright to Octave Chanute

December 23, 1901

As to your suggstion in regard to Mr. Carnegie, of course nothing would give me greater pleasure than to devote my entire time to scientific investigations; and a salary of ten or twenty thousand a year would be no insuperable objection, but I think it possible that Andrew is too hard-headed a Scot to become interested in such a visionary pursuit as flying.

I return the Langley letter which I have read with regret. I was not unprepared for his decision as some remarks Mr. Huffaker made last summer gave the impression that Lilienthal was not in high favor among the Washington group of workers for some reason. Mr. Huffaker seemed to think that Lilienthal had been overestimated. Since seeing his book I cannot help thinking that he is underestimated, and that he will stand even higher when the doubts with which some of his most important discoveries have been accepted are finally cleared away. It seemed to me that the publication of his book in English would not only be of very great value to all aeronautical workers but it would be a well deserved tribute to the memory of a man who spent much money, and immense amount of his time, and finally his life in carrying out investigations which he gave freely to the world.

*Dan Tate and Wilbur fly
the 1902 glider as a kite,
September 19, 1902.*

4

But It Was Cornbread

In baiting a mousetrap with cheese,
always leave room for the mouse.
—H. H. Munro, *The Square Egg*

Charles Taylor opened the bicycle shop every morning at seven o'clock excepting Sundays. Wilbur and Orville arrived between eight and nine. They staggered the lunch hour so that the shop was never closed until six p.m. except on Saturday when it stayed open until nine. Charles worked in the back room machine shop and waited on customers unless they particularly asked for one of the brothers.

Upstairs in the office the air was freighted with arguments about charts and figures, and sheets of paper on the work table were covered with graphs. Wilbur and Chanute exchanged lengthy letters on the average of seven a month, all of them discussing aspects of flying.

Chanute wrote the Wrights January 1, 1902, that the directors of the Louisiana Purchase Exposition, set for 1904, proposed substantial prizes for the encouragement of flying, although most of the plans were for balloons since most people believed that heavier-than-air flying was impossible. Wilbur said that he doubted they would be interested.

As soon as the proposal was announced in the newspapers, hundreds of men began building machines in cellars, attics, and stables. Everybody, said Chanute, had the problem of flight solved except whether to use steam, electricity or water-powered motors to drive their craft. "Mule-power might give more ascensional force if properly applied." said he, "but I fear it would be too dangerous unless the mule wore pneumatic shoes."

Augustus Herring, born the same year as Wilbur, 1867, had studied mechanical engineering at Stevens Institute of Technology. He left before graduation because his thesis "The Flying Machine as a Mechanical Engineering Problem" was not accepted. He began gliding experiments, copying those of Lilienthal. He assisted Langley at the Smithsonian from June to November in 1896. He then worked for Chanute, making and flying gliders.

Herring applied for a patent for a heavier-than-air powered flying machine December 11, 1896. Instead of the required model he submitted an affidavit telling of his gliding experiences and work on light engines. Three photographs accompanied the affidavit: a two-foot elastic band-propelled model in flight, one of himself gliding, and one of a light, two-cylinder gas engine.

Although the Patent Office found nearly twenty claims not anticipated by the prior art, the Office rejected Herring's application because "no power-driven airplane has yet been raised into the air with the aeronaut or kept its course wholly detached from the earth for such considerable time as to constitute practical usefulness."

Herring worked on gasoline and steam engines for several years, but his only success came in 1898 when he made a hop of seventy-five feet in his full-size machine powered by compressed air. A fire in 1901 destroyed his shop and partly-completed machine.

Chanute wrote Wilbur in July 1902 that Herring was out of a job and wanted to build two machines that Chanute had earlier asked the Wrights to build. He also suggested that the brothers should take out patents or caveats. Wilbur answered that the only reason he had agreed to build the machines was for the good of the cause and that he did not relish the idea of building machines for some other person to risk his neck.

Chanute asked to send his own machines to Kitty Hawk with Herring or William Avery to make the tests. Wilbur was reluctant to have the men in camp. "It was our experience last year that my brother and myself, while alone or nearly so, could do more work in one week than in two weeks after Mr. Huffaker's arrival." Wilbur had no way of knowing then that he should have said, "Absolutely not." Nevertheless, they consented because of their friendship for Chanute. They also invited George Spratt to return because he worked so willingly and was congenial.

One man who was a strong competitor in the race to be the first to fly a heavier-than-air machine was Samuel Pierpont Langley, the sixty-eight-year old secretary of the Smithsonian Institution. As early as 1889 he experimented to acquire the art of flying by studying stuffed birds on a whirling table. He concluded from those studies that it took more power to make a stuffed bird soar than a live one required and said, "We cannot carry nearly the weights which Nature does to a given sustaining surface without a power much greater than she employs."

Early in the 1900's he built models of machines powered by rubber bands and then advanced to a steam-driven machine. In 1896 one of his four-winged aerodromes, which weighed about 25 pounds and had a 14-foot wingspread, catapulted from a houseboat in the Potomac and flew about 3,000 feet. Alexander Graham Bell, who took photographs of that flight said, "It seems to me that no one who was present on this interesting occasion could have failed to recognize that the practicability of mechanical flight has been demonstrated."

Writing of his experiments in the June 1897 issue of *McClure's*, Langley said that the machine could lift itself, but that he had learned that when birds soar in the air, they find something more than strength of wing necessary to fly. From this he theorized that something more than mechanical power was needed to make the machine fly, but the something was not clear to him. "The first difficulty seemed to be to make the initial flight in such conditions that the machine would not wreck itself at the outset in its descent."

Assistant Secretary of War Theodore Roosevelt hoped that a man-carrying flying machine could be developed for use in the Spanish-American War. President William McKinley, at Roosevelt's urging, requested an appropriation of $50,000 for that purpose. The War Department Board of Ordnance and Fortification, after an investigation

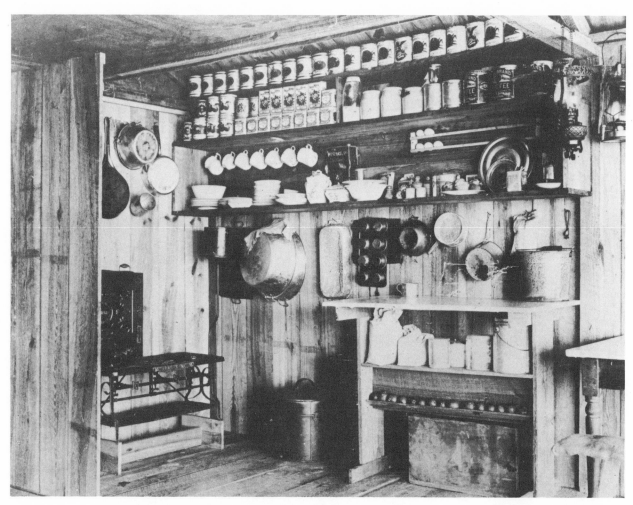

of Langley's experiments, granted him the money to develop the machine. The war ended before he could make much progress.

Wilbur wrote to Chanute in July 1902 that a family problem would prevent their departure for Kitty Hawk until late August. The bishop, who was seventy-three, had become involved in a church quarrel. A layman in the church publishing house at Huntington, Indiana, was suspected of mishandling church funds. Bishop Wright demanded that the man be removed, if not prosecuted, but several of the layman's friends wished to stop the scandal and also to remove the bishop from office. Wilbur went to Huntington to examine the publishing house books. When he returned, he reported to his father that the books were inaccurate.

Fifteen apples, 5 eggs, 8 cans cling peaches, 2 cans pineapple, 5 cans plums, 3 cans coffee, 5 tins sugar, 1 box cornstarch, 2 cans Royal Baking Powder, salt, pepper, spices, flour, cornmeal, tea, and cooking oil stand on the shelves in the 1902 Kitty Hawk camp.

In April Wilbur worked with his father preparing a paper revealing a financial discrepancy of $2,000. Orville typed the paper, and the bishop had it printed. He mailed 500 tracts to members of the United Brethren Church. The bookkeeper was brought to trial in early May. During the trial the bishop wrote another tract and mailed 2,000 copies. Wilbur spent a week with his father, traveling and making calls in his father's behalf. The bishop was brought to trial in a countersuit. Wilbur, as his father's defender, spent five days in Huntington in early August. The bishop was not removed, but the suit was not settled at the time.

Katharine to her father

(Wilbur and Orville) are talking of going next Monday though sometimes Will thinks he would like to stay and see what happens at Huntington next week. They really ought to get away for a while. Will is thin and nervous and so is Orv. They will be all right when they get down in the sand where the salt breezes blow. They insist that if you aren't well enough to stay out on your trip, you must come down with them. They think that life at Kitty Hawk cures all ills, you know.

The flying machine is in the process of making. Will spins the sewing machine around by the hour while Orv squats around, marking spaces to sew. There is no place in the house to live but I'll be lonesome this time next week and wish that I could have some of their racket around.

The brothers left Dayton August 25, 1902, by train for Elizabeth City, arriving at 5:45 p.m. the next day. At the wharf they found a schooner ready to leave for Kitty Hawk the next morning. They had fifteen minutes to collect the luggage from the baggage room before it closed and to buy a barrel of gasoline before the warehouse closed. Orville wheedled a hardware store manager into reopening his store long enough to sell him an oven.

The wind was too light for them to sail until 6:00 the following morning, and the trip took until late the next afternoon to tie up at Kitty Hawk. When they arrived at camp they found that the wind had blown the sand from under their building so that the floor sloped like a mountain side. They ate beef-extract soup and crackers and went early to bed.

Next day they drove a well, set up a dining table covered with two thicknesses of burlap and white oil cloth and as Wilbur wrote home, "We have upholstered our dining-room chairs with excelsior and burlap, and have put in other royal luxuries." They leveled and repaired the shed for their glider and built a smaller shed for a house. They built camp beds high up under the roof and set up an orderly and well-stocked kitchen.

Wilbur in 1902 glider with fixed double rudder over Kill Devil Hill, October 10, 1902

The fixed double rudder is clearly seen in this view of Wilbur flying the 1902 glider October 2, 1902.

Chanute told them he planned to send Herring or Avery to the Wright camp. "In a former letter," wrote Wilbur, "I expressed a preference for Mr. Avery because several things I heard about Mr. Herring's relations with Langley and yourself seemed to me to indicate that he might be somewhat of a jealous disposition, and possibly inclined to claim rather more credit for himself than those with whom he might be working would be willing to allow. If you should find it more convenient to send Mr. Herring it will be entirely satisfactory to us. If you also are in camp during the term that he is here I do not see how any misunderstanding will arise."

After they cleaned out the larger shed, the brothers began to put their glider together. The glider had a wingspan of 32 feet, a wing area of 305 square feet, a horizontal rudder area of 15 square feet, and a vertical double rudder with an area of 5.73 square feet. The glider was 16 feet long and weighed 112 pounds.

They began gliding September 20. By the end of the second day they had completed fifty manned glides with the machine acting exceptionally well.

Even the mosquitoes were not so bad as the year before, but another pest bothered them. "I put in part of the day in constructing a death trap for a poor mouse that had been annoying us by prowling around our kitchen shelves," Orville wrote in his diary September 26. "We are now awaiting anxiously the arrival of the victim."

Next day he continued, "At 11 o'clock last night I was awakened by the mouse crawling over my face. Will had advised me that I had better get something to cover my head or I would get it 'chawed' off. I found on getting up that the little fellow had only come to tell me to put another piece of cornbread in the trap. He had disposed of the first piece. I have sworn vengeance on the little fellow for this impudence and insult."

Lorin Wright came to camp September 30 and George Spratt arrived the next day. At breakfast the next morning Orville caught Lorin's eye to alert him that he was going to say something important. "While I was lying awake last night," Orville said, "I studied out a new vertical movable rudder to replace the fixed rudder we have used."

Orville expected that because Wilbur was the older brother, he would brush aside the idea or assume that he himself had thought of it, as he often did. This time Wilbur listened carefully to his younger brother. He saw immediately that Orville had hit on as revolutionary an idea as grew from his own twisting of the bicycle box.

The movable single vertical rudder or tail fin is the fourth major Wright invention. It enabled the operator to control the yaw, the direction of the flight path.

The modified 1902 glider had for the first time control over all three flight motions: wing warping for roll control, elevator for pitch control, and a single vertical movable rudder for yaw control.

Wilbur saw at once that Orville's idea would solve a long-standing problem.

When a glider flew forward and tilted to one side, the speed increased and if the operator did not correct the balance quickly, the machine slid faster and the wind struck the side toward the low wing rather than the high wing as it was expected to do. The vanes of the rudder did not turn the glider but in fact made it turn the wrong way even faster with results referred to in later years as a tailspin.

"And I know how we can make it even better," Wilbur said after hearing Orville's explanation of how a movable rudder would keep the craft level in the lateral plane. "We will connect the rudder and the wings with wires so as to operate them simultaneously. One lever will control lateral balance. The other lever will keep the machine balanced fore and aft."

They started building the new rudder the next day, and as Wilbur was to say in court in 1912, "After the adjustable rudder was installed not once did we encounter the difficulty we had experienced with the fixed vane."

Chanute and Herring arrived at Kitty Hawk October 5. The next day Herring put the multiple-wing glider together, and Orville found the mouse dead under a trunk.

Several days of flights followed. The Wright glider flew successfully but the Chanute plane did not. Herring said it was useless to continue with Chanute's glider. Orville said, not in Chanute's hearing, "I think that much of the trouble is in the glider's structural weakness. In winds not even strong enough for support the surfaces are distorted so that while the wind at one end is on the underside, often at the other end it is on top."

Lorin left for home October 13. Orville, Spratt, and Herring set up Chanute's second plane. Chanute made one glide with it as a kite, but it flew only 50 feet.

Tuesday night, October 14, Herring wakened the entire camp. "A fox has stolen our chicken," he announced. Since he had made similar announcements several times before, nobody paid any attention. When daylight came, they found the chicken safe in the coop.

Chanute and Herring took the glider out once more, but Chanute realized the measurements were wrong and said to take it back inside the shed. That same afternoon Chanute and Herring left camp.

Chanute's three-winged glider with Augustus Herring hanging by his arms; Wilbur, left; Dan Tate, right, October 13, 1902

Orville in glider, Wilbur, left; Dan Tate, right, early October 1902

After the two had gone, the Wrights had five perfect days of gliding. "Day before yesterday," Orville wrote Katharine, "we had a wind of about 16 meters per second or about 30 miles an hour and glided in it without any trouble. That was the highest wind a gliding machine was ever in, so that now we hold all the records!" The records were for handling the largest machine in any kind of weather, the longest American distance glide, the longest time in the air, and the smallest angle of descent in the highest wind. They had increased their record for distance to 622½ feet, for time to 26 seconds, and for angle of descent to 5 degrees for 156 feet.

Octave Chanute to Samuel P. Langley

October 21, 1902

I have lately gotten out of conceit with Mr. Herring, and I fear that he is a bungler. He came to me in July, said that he was out of employment and urged that I let him build the gliding machines, "to beat Mr. Wright." I consented to building new wings for the multiple-wing machine, but could give it no attention as the work was done at St. Joseph, Mich. Herring adopted new forms of wings and reduced the total weight from 33½ lbs. to 27 lbs., but when the machine was tried by him in N. C. it proved a failure, and he said he did not know what was the matter.

Samuel P. Langley to Chanute

October 23, 1902

I had already made up my mind about the gentleman in question. After seeing you I almost decided to go or send someone to see the remarkable experiments that you told me of by the Wright brothers. I telegraphed and wrote to them at Kitty Hawk but had no answer, and I suppose their experiments are over.

When the Wrights received Langley's letter and telegram, they answered that as they were leaving in a few days, it was impossible to invite Langley to camp. They broke camp and arrived back in Dayton October 31.

Chanute told the brothers that he was going to make a trip to Egypt. Wilbur said it was a pleasure to learn about the trip and that he supposed Chanute would enjoy it very much. "It is a land of wondrous interest," he said, "and the home of many remarkable birds whose evolutions will doubtless share your attention along with the Pyramids, Thebes, and the great dam."

In December 1902 the brothers sent letters of inquiry to manufacturers of gasoline engines, asking for an engine that would develop 8 to 9 horsepower, weigh no more than 180 pounds, and be free from vibration. Ten companies replied that they did not make such an engine. "Well," said Wilbur to Charles Taylor, "we'll just have to build our own."

"I suppose I don't have to ask you why you want it."

"That's right. We are going to put it on our machine. Next year at Kitty Hawk we are going to fly."

"A remarkable picture," says historian Paul E. Garber, "because it documents the realization of control." Until October 24, 1902, no glider had three-axis control: roll, pitch, and yaw. Wilbur banks to the right, using the new single, movable rudder attached by cables to the hip cradle, which controlled wing warping.

5

Screw Your Courage

But screw your courage to the sticking-place,
And we'll not fail.
　　—Shakespeare, *Macbeth, I, vii, 59*

"They figured on four cylinders and estimated the bore and stroke at four inches," reminisced Charles Taylor in later years:

While the boys were handy with tools, they had never done much machine work and anyway they were busy on the air frame. It was up to me. My only experience with a gasoline engine was an attempt to repair one in an automobile in 1901.

We didn't make any drawings. One of us would sketch out the part we were talking about on a piece of scratch paper and I'd spike the sketch over my bench.

It took me six weeks to make that engine. The only metal-working machines we had were a lathe and a drill press, run by belts from the stationary gas engine.

The crankshaft was made out of a block of machine steel 6 by 31 inches and one and five-eighths inches thick. I traced the outline on the slab, then drilled through with the drill press until I could knock out the surplus pieces with a hammer and a chisel. Then I put it in the lathe and turned it down to size and smoothness. It weighed 19 pounds finished and she balanced up perfectly.

The completed engine weighed 180 pounds and developed 12 horsepower at 1,025 revolutions per minute.

The body of the 1903 engine was of cast aluminum, and was bored out on the lathe for independent cylinders. The pistons were cast iron, and these were turned down and grooved for piston rings. The rings were cast iron, too.

The fuel system was simple. A one-gallon fuel tank was suspended from a wing strut, and the gasoline fed by gravity down to the engine. The fuel valve was an ordinary gaslight petcock. There was no carburetor as we know it today. The fuel was fed into a shallow chamber in the manifold. Raw gas blended with air in this chamber, which was next to the cylinders and heated up rather quickly, thus helping to vaporize the mixture. The engine was started by priming each cylinder with a few drops of raw gas.

Wilbur and Orville designed it; Charlie Taylor built it in the bicycle shop—the 1903 engine.

The ignition was the make-and-break type. No spark plugs. The spark was made by the opening and closing of two contact points inside the combustion chamber. These were operated by shafts and cams geared to the main camshaft. The ignition switch was an ordinary single-throw knife switch we bought at the hardware store. Dry batteries were used for starting the engine and then we switched onto a magneto bought from the Dayton Electric Company. There was no battery on the plane.

Several lengths of speaking tube, such as you find in apartment houses, were used in the radiator.

A simple chain and sprocket wheel drove each of the two propellers. The wheel was attached to the propeller shaft; the chain ran down to the motor.

The chains were specially made by the Indianapolis Chain Company, but the sprockets came ready-made. Roebling wire was used for the trusses.

I think the hardest job Will and Orv had was with the propellers. I don't believe they were ever given enough credit for that development. They had read up on all that was published about boat propellers, but they couldn't find any formula for what they needed. So they had to develop their own, and this they did in the wind tunnel.

They concluded that an air propeller was really just a rotating wing, and by experimenting in the wind box they arrived at the design they wanted. They made the propellers out of three lengths of wood, glued together at staggered intervals. Then they cut them down to the right size and shape with a hatchet and drawshave. They were good propellers.

We never did assemble the whole machine at Dayton. There wasn't room enough in the shop. When the center section was assembled, it blocked the passage between the front and back rooms, and the boys had to go out the side door and around to the front to wait on the customers.

We still had bicycle customers. The Wright brothers had to keep that business going to pay for the flying experiments. There wasn't any other money.

While the boys always worked hard, and there never was any horseplay around the shop, they always seemed to find time to stop and talk with a customer or humor the neighborhood children who wandered in. Sometimes I think the kids were the only ones

The propeller is the fifth great Wright invention.

Historian emeritus Paul E. Garber, the National Air and Space Museum, the Smithsonian Institution, says, "Their work resulted in the development of aircraft propellers having 66 percent efficiency, which had never been attained by any other inventor."

who really believed that Will and Orv would fly. They hadn't learned enough to say it couldn't be done.

We block-tested the motor before crating it for shipment to Kitty Hawk. We rigged up a resistance fan with blades an inch and a half wide and five feet two inches long. The boys figured out the horsepower by counting the revolutions per minute. Those two sure knew their physics. I guess that's why they always knew what they were doing and hardly ever guessed at anything.

We finally got everything crated and on the train. There was no ceremony about it, even among ourselves. The boys had been making these trips for four years, and this was the third time I had been left to run the shop. If there was any worry about the flying machine not working, they never showed it and I never felt it.

"For two reasons we decided to use two propellers," Orville would write in December, 1913. "In the first place we could, by the use of two propellers, secure a reaction against a greater quantity of air, and at the same time use a larger pitch angle than was possible with one propeller: and in the second place, by having the propellers turn in opposite directions, the gyroscopic action of one would neutralize that of the other.

"We decided to place the motor to one side of the man, so that in case of a plunge headfirst, the motor could not fall upon him. In our gliding experiments we had had a number of experiences in which we had landed upon one wing, but the crushing of the wing had absorbed the shock, so that we were not uneasy about the motor in case of a landing of that kind. To provide against the machine rolling over forward in landing, we designed skids like sled runners, extending out in front of the main surfaces. Otherwise the general construction and operation of the machine was to be similar to that of the 1902 glider.

"When the motor was completed and tested, we found that it would develop 16 horsepower for a few seconds, but that the power rapidly dropped till, at the end of a minute, it was only 12 horsepower. Ignorant of what a motor of this size ought to develop, we were greatly pleased with its performance. More experience showed us that we did not get one-half of the power we should have had.

"With 12 horsepower at our command, we considered we could permit the weight of the machine with operator to rise to 750 or 800 pounds, and still have as much surplus power as we had originally allowed for in the first estimate of 550 pounds.

"Before leaving for our camp at Kitty Hawk, we tested the chain drive for the propellers in our shop at Dayton, and found it satisfactory."

Chanute to Wilbur

Paris, April 4, 1903

Your experiments are attracting a good deal of attention in Paris. *L'Aerophile* wants your picture and that of your brother Orville to publish with the article I have agreed to prepare. You are therefore upon the receipt of this to go to the photographer and be "took," and to send me two copies of each at Chicago, where I expect to be on May 8th to 10th.

Wilbur wrote back that Chanute's request for photographs was causing them a great deal of mental distress. "We do not know how to refuse you when you have put the matter so nicely, and on the other hand, we haven't the courage to face the machine. While we were waiting to get our courage screwed up to the sticking point, Orville managed one day to get a grain of emery in his eye. It caused quite a severe inflammation and compelled him to remain in a darkened room one day, but it is getting better now."

Propeller
1903

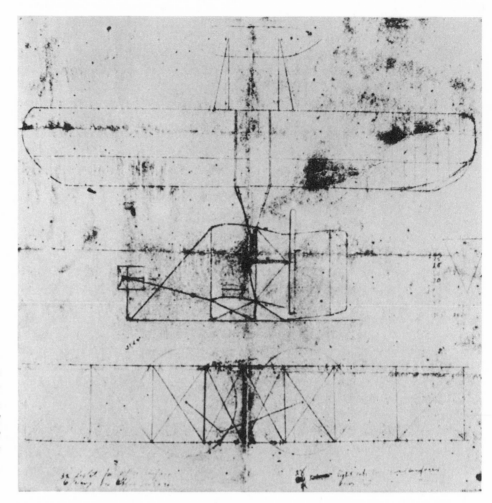

Wilbur drew on brown wrapping paper this preliminary sketch showing top, side, and front view of the 1903 machine. The original sketch is in the Franklin Institute, Philadelphia.

The brothers never did send their photographs, writing later to Chanute, "Really, we would rather not."

George Spratt had been writing to Wilbur over several months about several flying problems and asking Wilbur's advice. When Spratt expressed discouragement, Wilbur wrote to him April 20, 1903, "Discussion brings out new ways of looking at things and helps to round off the corners. You make a great mistake in envying me any of my qualities. Very often what you take for some special quality of mind is merely facility arising from constant practice, and you could do as well or better with like practice. It is a characteristic of all our family to be able to see the weak points of anything, but this is not always a desirable quality as it makes us too conservative for successful business men, and limits our friendships to a very limited circle. You envy me, but I envy you the possession of some qualities that I would give a great deal to possess in equal degree.

"Regarding the matters on which you have asked my advice I must confess that I am at a loss just what to say. Regarding a matter which might affect the whole course of a man's life, I almost fear to give any advice, lest injury might result from it, instead of good, as intended. I can suffer the consequences of my own mistakes with some composure, but I would hate awfully to see some other person suffering from an error of judgment of mine."

After further exchange of several somewhat argumentative letters between Wilbur and Spratt, Orville wrote to Spratt in June, 1903. He said, "Wilbur and I both take great interest in your letters, and my not writing to you is not from a lack of interest in what you are doing, but rather from lack of ability as a letter writer. Will seems to enjoy writing, so I leave all the literary part of our work to him."

He then described what they had been doing since Spratt and they had parted at Kitty Hawk and ended his letter by saying:

During time the engine was building we were engaged in some very heated discussions on the principles of screw propellers. We had been unable to find anything of value in any of the works to which we had access, so that we worked out a theory of our own on the subject, and soon discovered, as we usually do, that all propellers built heretofore are *all wrong,* and then built a pair of propellers 8½ feet in diameter, based on our theory, which are *all right!* (till we have a chance to test them down at Kitty Hawk and find out differently). Isn't it astonishing that all these secrets have been preserved for so many years just so we could discover them! Well, our propellers are so different from any that have been used before that they will either have to be a good deal better, or a good deal worse.

P.S. Please do not mention the fact of our building a power machine to anybody. The newspapers would take great delight in following us to record our troubles.

A month later Chanute asked Wilbur whether he had heard that Langley was about to test his man-carrying machine. "Prof. Langley seems to be having more than his fair share of trouble just now with pestiferous reporters and windstorms," Wilbur answered. "But as the mosquitoes are reported to be very bad along the banks where the reporters are encamped he has some consolation. It would be interesting to attempt a computation of the possible performance of his machine in advance of its trial, but the data of the machine as given in the newspapers are evidently erroneous that it seems hopeless to attempt it. It is a sure thing that the speed will not be from 60 to 90 miles an hour with an expenditure of 25 horsepower as the papers have reported its prospective flight. I presume that you are to be one of the guests of honor at the launching festivities. Our invitation has not yet arrived."

Wilbur had to make another trip to Huntington to help his father with the church problem before the brothers could leave for Kitty Hawk September 23, 1903. Orville wrote to Katharine from Kitty Hawk:

We had supposed two years ago, when the wind speed of 107 miles per hour took the anemometer cups away with it (beating anything within the memory of the oldest inhabitants), and when the mosquitoes were so thick as to dim the very brightness of the sun, exceeding in numbers all excepting those that devoured the whole of Raleigh's settlers on Roanoke, and last year when the lightning turned night into day, and burned down every telephone pole between here and Kitty Hawk, we had supposed that Nature had reached her limit; but far from it! Dan says this year has been one continuous succession of storms of unprecedented severity; the rain has descended in such torrents as to make a lake for miles about our camp; the mosquitoes were so thick they turned day into night, and the lightning so terrible that it turned night into day. Really it paralyzes the mind to try to think of all these things at once. Besides all those, the sun was so hot, it must have made soup out of the mosquitoes and rain!

Orville spent most of one rainy morning inventing a French drip coffee pot in order to do away with boiled coffee that had to be cleared with egg shells. It was a good coffee pot except that it had a way of boiling over and mixing the coffee grounds and water. He told Katharine to have Lorin get some of the very finest mesh wire brass screen and send it down immediately for the coffee pot.

Since the old building had been blown off its foundation during the winter, the brothers put up a new one, 44 by 16 by 9 feet.

On October 14 they received a copy of *The Washington Post* of October 8, 1903. Samuel P. Langley's man-carrying powered heavier-than-air flying machine had been placed atop a houseboat on the Potomac River. The machine was a tandemwing aerodrome with a 53 horsepower gasoline, five-cylinder radial engine built by Langley's assistant, Charles Manley.

The Washington Post, Oct. 8—A few yards from the houseboat were the boats of the reporters, who for three months had been stationed at wide water (along the Potomac). The newspapermen waved their hands. Manley (the pilot) looked down and smiled. Then his face hardened as he braced himself for the flight, which might have in store for him fame or death. The propeller wheels, a foot from his head, whirred about him 1,000 times a minute. A man forward fired two sky rockets.

There came an answering "Toot, toot" from the tugs. A mechanic stooped, cut the cable holding the catapult; there was a roaring, grinding noise—and the Langley airship tumbled over the edge of the houseboat and disappeared in the river, 16 feet below. It simply slid into the water like a handful of mortar.

A gale hit Kill Devil Hills October 15, lasting three days. Dan Tate told the brothers that it broke all records for persistence. Five vessels were driven ashore between Kitty Hawk and Cape Henry. Part of the roof of the building gave way and Orville, fearing that the whole roof would go if he didn't do something to stop it, put on Wilbur's heavy overcoat, grabbed a ladder and mounted to the edge of the roof. "The wind caught under his coat and folded it back over his head," wrote Wilbur. "As the hammer and nails were in his pocket and up over his head he was unable to get his hands on them or to pull his coattails down, so he was compelled to descend again. The next time he put his nails in his mouth and took the hammer in his hand and I followed him up the ladder hanging on to his coattails. He swatted around a good little while trying to get a few nails in, and I became almost impatient for I had only my common coat on and was getting well soaked. He explained afterward that the wind kept blowing the hammer around so that three licks out of four hit the roof or his fingers instead of the nail. The wind and rain continued through the night, but we took the advice of the Oberlin coach, 'Cheer up, boys, there is no hope.'"

They went to bed. In the morning they found most of the floor under water, but their kitchen and dining room were all right.

Wilbur to Chanute

Kill Devil Hills, October 16, 1903

We regret to learn there is danger (possibility) of your being unable to visit our camp this year. We are expecting the most interesting results of any of our seasons of experiment, and are sure that, barring exasperating little accidents or some mishap, we will have done something before we break camp.

The upper surface of the new machine is completed. It is far ahead of anything we have built before.

I see that Langley has had his fling, and failed. It seems to be our turn to throw now, and I wonder what our luck will be. We still hope to see you before we break camp.

The 1903 machine, the shed, and the cabin at Kitty Hawk November 24, 1903

After the storm ended and mail could be delivered, the brothers received a newspaper clipping from Chanute:

Washington, Oct. 24—The Army Board of Ordnance and Fortification, when it meets on November 8, will take up the question of allotting additional funds to Professor S. P. Langley for his plan of mechanical flight. Professor Langley will make to the Board a report of his recent experiment.

In the note that accompanied the newspaper clipping Chanute said he thought perhaps he would be present at the trials at Kitty Hawk, and that he was surprised that the newspapers had not yet spotted them.

While they were working on their big machine, the brothers made more than 200 manned glides in the 1902 machine for flying practice. In one glide, Orville noted in his diary, "a strong gust started to take me up rapidly, and in bringing it down I lit upon Will's head with the rear left-hand corner." Will was not hurt, but both front and rear upper spars were broken.

Excerpts from Orville Wright's Diary D, 1903

October 23—Worked on skids during morning, and after dinner finished putting on hinges. Dr. Spratt arrived. Sat up until 11:00 discussing some of his theories of flight.

October 24—We put in the uprights between the surfaces and trussed the center section. Had much trouble with wires, which failed to fit.

October 28—Spratt and I passed a very miserable night on account of cold. We worked today on front rudder frame and uprights between rudder surfaces. We spent afternoon in making a stove and in putting it up. Dan struck on account of having to bring in wood for stove and left at 3 o'clock.

October 30—Continued work on front rudder and completed it this evening. Took machine out and turned it about ready for putting on tail in morning. Weighed machine, with wires, and all connections, but minus tail and front rudder frames and surfaces, which amounted to 227 lbs. Front rudder complete weighs 29 lbs.

Orville to Katharine

Kitty Hawk, November 1, 1903

I suppose you have read in the papers the account of the failure of Langley's big machine. He started from a point 60 feet in the air and landed 300 feet away, which is a drop of 1 foot for every 5 forward. We are able, from this same height, to make from

400 to 600 feet without any motor at all, so I think his surfaces must be very inefficient. They found they had no control of the machine whatever, though the wind blew but 5 miles an hour at the time of the test. That is the point where we have a great advantage. We have been in the air hundreds and hundreds of times, and have pretty well worked out the problem of control. We find it much more difficult to manage the machine in one spot than when traveling rapidly forward. We expect no trouble from our big machine at all in this respect. Of course we are going to thoroughly test the control of it before attaching the motor. We are highly pleased with our progress so far this year. I have been putting in about an hour every night in studying German.

Excerpts from Orville's Diary D, 1903

November 2—Began work of placing engine on machine, also uprights for screws.

November 4—Have machine now within half day of completion. Spratt made track 60 ft. in length.

November 5—The wind during the day has been blowing from the southwest with a velocity of 6 to 8 meters. We got the machine ready for testing engine and screws (propellers) in building preparatory to taking it out. We had much trouble in fastening the sprockets on tight, and found finally that the lock-nut screwed up against shaft without touching sprocket at all. This allowed the screws to play almost a half turn. After a number of trials to get the engine running properly, the pieces to which the screws were fastened were jerked loose from both shafts, thus taking away all chance of a trial for ten days or so. We had found, also, that the magneto, which we were running in the opposite direction to what it has been before, failed to produce a spark sufficient to ignite the gas. Spratt, seeing that the machine could not be ready for some time, decided to leave at once, and left about 4 o'clock. We sent with him the shafts to be expressed from Norfolk, also letter to be mailed giving Charlie instructions as to repairs.

November 6—Capt. Midgett's boy brought down groceries this morning and announced that Mr. Dosher had a telegram for us from Mr. Chanute who was on his way to visit us. The weather is very cold, and a light rain has been falling most of the day. After dinner Mr. Etheridge of Life Saving Sta. came over through rain with telegram from Chanute. While warming at our stove, his horse pulled stake to which he was tied loose and after scurrying about for a while ran off to the station leaving Mr. Etheridge to walk back. Mr. Chanute arrived about four o'clock.

November 8—Sat up to 11 o'clock talking over with Mr. Chanute plans for next year, should machine make a successful flight.

November 9—Spent most of day working on engine and magneto, which after much trouble was got into shape for running. The vibration at high speed is not at all disagreeable.

November 11—In afternoon we put machine on track and ran it out of building to finish some of fastenings of front rudder. Mr. Chanute went to Station to make arrangements for going to Manteo tomorrow morning.

November 12—Mr. Chanute left with Mr. Dough of the Kill Devil Station in his sailboat for Manteo at eight o'clock. On our return to camp we began work in planing down the starting track. We decided to test our method of starting from track with old machine,

Wilbur and Orville assemble the 1903 machine.

so we took two rails to Big Hill. Five starts out of six were successful. We then tried starting with one man holding front rudder frame, and making a few steps backward. After four or five flights we took machine back to camp. Spent most of afternoon in chopping wood and reading.

November 13—The wind being very quiet, we took new machine out to add braces to skids and to attach wires to front rudder frame.

November 14—We took front rudder out to measure pressure on surfaces, which appeared to be about 22 lbs. on one side and about 40 lbs. on the other. In afternoon we took it out again in a wind of 15 meters to test its strength. It stood up all right when surfaces were turned to their greatest angles. Spent evening making calculations on our screws. We find that our gears are too high so that we cannot use enough of the power of the engine in starting. However we are still hopeful. The engine will have to make 875 rev. per min. (305 of screw) to furnish 100 lbs. thrust standing still. If it fails to do this there will be little hope of success until we get new gears so as to speed engine up.

Orville to Bishop Wright and Katharine

November 15, 1903

The weight of our machine complete with man will be a little over 700 lbs. and we are now quite in doubt as to whether the engine will be able to pull it at all with the present gears, as we will not be able to use more than 3/4ths of our power in getting started. The screws came loose before we had time to either measure the speed of the engine or the thrust of the screws. Mr. Chanute says that no one before has ever tried to build a machine on such close margins as we have done to our calculations. He said that he nevertheless had more hope of our machine going than any of the others. He seems to think we are pursued by a blind fate from which we are unable to escape.

The past week and a half has just been a loaf, since we have almost nothing to do on the machine until the shafts come. I am taking up my German and French again, and am making some progress. An article from a German paper, giving some account of our machine, has kept me guessing, as I have only been able to find a few words in the vocabulary I have.

Excerpts from Orville Wright's Diary D, 1903

November 19—On arising found pools about camp frozen, also water in basin. Coldest night we have had so far. Gathered several logs of oak for firewood. Wind blowing 6 to 8 meters. Too cold for work.

November 20—Last night was not so cold. Wind still from north blowing 10 to 13 meters all day. We collected some more firewood before noon. Capt. Midgett brought our groceries and propeller shafts down about noon. After dinner we put them in place and toward evening we were ready to test. Had trouble with magneto which failed to furnish spark enough and could not get the sprockets to stay tight on our propeller shafts. The power transmission weighs as follows:

Propeller shafts, sprockets and frames, 18½ & 20, together . . . . 38½ lbs.
Long chain with guides . . . . 24 lbs.
Short chain with guides . . . . 14½ lbs.
Propellers (together) . . . . 13 lbs.

Day closes in deep gloom.

November 21—After many attempts to fasten sprockets we finally succeeded by filling thread with tire cement. The engine ran very irregularly, jerking the chains, and shaking the machine terribly. We discovered the trouble lay in the gasoline feed, and, after fixing valve so that the vibration could not change it, we had no further trouble from that source. The first test of speed was 306 rev. screw to minute (309 in 60½ sec.). One cylinder made only a few explosions during the test. On the next trial we got 333 rev. in 60 sec. After dinner we arranged to measure the thrust by supporting center skids on rollers and fastening one end of the machine, while we attached a rope to the other end, which ran over a pulley and carried a 50 lb. box of sand. Besides lifting the sand we got an additional pull of 16 to 18 lbs. on the scales, which made the total thrust of the screws 132 to 136 lbs. at a speed of 350 revolutions per min. Our confidence in the success of the machine is now greater than ever before.

November 23—Completed rails for starting machine. Found that truck loaded to 270 lbs. would run down slope of 1°5'. Each took two trips with track laid on 11° slope of Big Hill. Time 4 and 2/5 sec. Attached wires to cradle for operating tips. Pressure required for changing tips 12° is 14 lbs. Tested amt. of gasoline used when running screws at 350, which was found to be 1 gal. in 45 min. or the amount held by tank, 3 lbs. 3 oz., in 18 min.

The starting track was built of four 15-foot two-by-fours, topped with a metal strip. The truck, which was used to support the skids, had two parts: a crosspiece shaped on top like a yoke and two tandem rollers, made of bicycle hubs, which moved along the track on the metal strip.

Wilbur to Bishop Wright and Katharine

Kitty Hawk, November 23, 1903

Our track for starting the machine (total cost about $4) amused Mr. Chanute considerably, as Langley is said to have spent nearly $50,000 on his starting device, which failed in the end to give a proper start he claims. At least this is the reason he gives for the failure last month. We have only tried ours with the little machine so far, but it seems to work well.

In a November 23 letter to Charles Taylor congratulating him on the excellent repair on the shafts, Orville said, "We have not set the date of our return yet. It will depend on circumstances. However it will not be before the 8th or 10th of December. We will try to get home for Christmas."

6

The Untraveled World

Yet all experience is an arch wherethrough
Gleams that untravel'd world.
—Tennyson, *Ulysses*

Orville Wright went home alone three weeks before Christmas.

On the morning of Saturday, November 28, when the brothers were testing the engine, they discovered that one of the propeller shafts had cracked. "We don't have time," Orville said, "to sit around here waiting while those shafts go by express and winter closes in on us. I'm going home tomorrow to make new shafts of solid tool steel."

"Tomorrow is Sunday," said Wilbur.

Orville left for Dayton on Monday, November 30.

While Orville was at home making the new shafts, Langley supervised the second attempt for his aerodrome to fly on December 8. The engine started and with Manley at the controls began to go down the launching track. "This time the rear of the engine," as Langley was to write later in *Experiments with the Langley Aerodrome,* "in some way still unexplained, was caught by a portion of the launching car which caused the rear sustaining surfaces to break, leaving the rear entirely without support, and it came down almost vertically into the water."

Langley's machine broke into two parts and sank into the Potomac. A quick dive into the river by one of Manley's colleagues saved Manley from drowning. The press wrote derisively of Langley's second failure. The failure was responsible to a great degree for the public's belief that heavier-than-air flight was impossible.

Orville arrived at camp December 11 with the new shafts.

Excerpts from Orville Wright's Diary D, 1903

December 12—Set propeller shafts and got machine outside in afternoon with intention of making a trial. We did not have enough wind for starting from flat and not enough time to go to the hill. We spent some time in running machine along track to see what

Landing after first launch, December 14, 1903; Wilbur at controls of damaged Flyer

speed one man could give it. In a 40-ft. run the last 15 feet were covered in 1½ sec. In starting one time the frames supporting the tail were caught on the track and broken.

Sunday, December 13—Spent most of day reading. In afternoon Mr. Etheridge of L. S. Station, with wife and children, called to take a look at machine.

December 14—We spent morning in making repairs on tail, and truck for starting. At half past one o'clock we put out signal for station men, and started for hill which took us about 40 minutes. After testing machine, with help of men (Bob Wescott, John T. Daniels, Tom Beacham, W. S. Dough, and Uncle Benny O'Neal), we took machine 150 ft. uphill and laid track on 8° 50' slope. A couple small boys, who had come with the men from the station, made a hurried departure over the hill for home on hearing the engine start. We tossed up coin to decide who should make first trial, and Will won. After getting adjustments of engine ready I took right end of machine. Will got on. When all was ready Will attempted to release fastening to rail, but the pressure due to weight of machine and thrust of screws was so great that he could not get it loose. We had to get a couple of the men to help push machine back till rope was slipped loose. While I was signaling man at other end to leave go, but before I myself was ready, Will started machine. I grabbed the upright the best I could and off we went. By the time we had reached the last quarter of the third rail (about 35 to 40 feet) the speed was so great I could stay with it no longer. I snapped watch as machine passed end of track. (It had raised from track six or eight feet from end.) The machine turned up in front and rose to a height of about 15 feet from ground at a point somewhere in neighborhood of 60 feet from end of track. After thus losing most of its headway it gradually sank to ground turned up at an angle of probably 20° incidence. The left wing was lower than the right so that in landing it struck first. The machine swung around and scraped the front skids so deep in sand that one was broken, and twisted around until the main strut and brace were also broken, besides the rear spar to lower surface of front rudder. Will forgot to shut off engine for some time, so that record of screw turns was mostly taken while the machine was on the ground. The engine made 602 revs. in 35½ s. Time of flight from end of track was 3½ sec. for a distance of 105 ft. Speed of wind was between 4 and 8 miles.

"The real trouble," Wilbur explained to his father and sister in a December 14 letter, "was an error in judgment, in turning up too suddenly after leaving the track, and as the machine had barely speed enough for support already, this slowed it down so much that before I could correct the error, the machine began to come down, though turned up at a big angle. Toward the end it began to speed up again but it was too late, and it struck the ground while moving a little to one side, due to wind and a rather bad start. It was a nice easy landing for the operator. The machinery all worked in entirely satisfactory manner, and seems reliable. The power is ample, and but for a trifling error due to lack of experience with this machine and this method of starting the machine would undoubtedly have flown beautifully. There is now no question of final success."

The brothers spent December 15 repairing the front rudder and rudder frame. It took until noon on December 15 to complete the repairs. They placed the machine on the track, planning to make a trial from level ground. While they were working, making the final adjustments, a stranger walked up. He stood watching them for a few seconds. He asked, "What is that?"

"It is a flying machine."

"Do you intend to fly it?"

"Yes, we do as soon as we have a suitable wind."

The stranger stood watching for several minutes longer and remarked, "It looks as if it will fly if you have a suitable wind." He walked away.

Wilbur said, "No doubt he's thinking of a 75-mile-an-hour gale we had the other day when he repeated our words 'a suitable wind.'"

By the time they were ready the wind had dropped to between four and five meters per second. They waited several hours but the breeze never returned and they put the machine away.

Thursday, December 17, 8 a.m.

Wilbur and Orville climbed down out of their ceiling-height beds, heated water on the carbide stove, and shaved. Wilbur was thirty-six years old and stood five feet ten and a half inches tall and weighed a little more than 140 pounds. Orville, who was thirty-two, stood an inch and a quarter shorter than Wilbur but weighed the same. Both had blue-gray eyes and brown hair. Wilbur's was darker than Orville's and had receded from his forehead and along the part. Orville's thicker hair had a reddish tinge and a slight curl.

They dressed in three-piece dark business suits, white shirts with attached standing collars, four-in-hand ties, dark hose and dark shiny city shoes.

8:30 a.m.

They ate their breakfast of biscuits and coffee, washed and wiped the dishes, and put them away. They stepped outside the shed. The puddles of water, which had been standing around the camp since the recent rain, lay covered with ice. They checked the wind velocity. "It's 10 to 12 meters per second," said Orville, looking at the anemometer.

"I hope the wind will die down after a while," said Wilbur. "Let's go back inside."

This view camera, bought by the Wright brothers in 1898, is on display in Carillon Park, Dayton. The camera has one of the first anastigmatic lenses.

*Front view of the
1903 Flyer*

10 a.m.

The brothers stepped outside again and read the anemometer. "It's just the same," said Wilbur. "Do you think we should get the machine out?"

"I think if we face the machine into the wind, there ought to be no trouble launching it from level ground," Orville answered. "Let's hang out the flag."

They hoisted the signal flag for the men at the Life Saving Station.

Orville and Wilbur laid the track on a smooth stretch of sand about 100 feet north of the glider shed. It was so cold that every few minutes they had to go inside the shed to warm up beside the carbide stove.

10:20 a.m.

John T. Daniels, W. S. Dough, and A. D. Etheridge, members of the Kill Devil Life Saving Station, W. C. Brinkley of Manteo, and Johnny Moore, a boy from Nag's Head, arrived. The men helped Wilbur and Orville carry the machine from the shed and set it on the rail. The brothers took a wind measurement; the anemometer showed a velocity of 11 to 12 meters per second.

10:27 a.m.

"Mr. Daniels, will you come with me," said Orville, as he picked up the camera, which had been placed on its tripod leaning against the building. "I'm going to set up the camera for you to take a picture." The two men walked to a spot between the camp buildings and the machine. Orville pushed the legs of the tripod into the sand and opened the camera, pulled out the bellows, set the exposure, and aimed it at a point a few feet short of the end of the starting rail. He pulled out the slide and handed it to Daniels. "Hold this in your left hand. As soon as the machine gets to the end of the rail, you squeeze the bulb. Then put the slide back in right here."

Orville walked back and drew Wilbur aside out of hearing of the other men. They talked briefly and shook hands. They walked to the machine and started the engine. The anemometer read 11 to 12 meters per second.

10:35 a.m.

While the engine was warming up, Orville climbed into the machine and lay prone in the operator's cradle. He pulled the visor of his cap over his forehead and put his hands on the controls. Wilbur walked to the men and said, "Don't look sad. Laugh and clap your hands and try to cheer Orv when he starts."

Wilbur took his position by the wing tip to steady it.

"On slipping the rope the machine started off increasing in speed to probably 7 or 8 miles," Orville wrote in his December 17 diary. "The machine lifted from the truck just as it was entering on the fourth rail. Mr. Daniels took a picture just as it left the track. I found the control of the front rudder quite difficult on account of its being balanced too near the center and thus had a tendency to turn itself when started so that the rudder was turned too far to one side and then too far on the other. As a result the machine would rise suddenly to about 10 ft. and then as suddenly, on turning the rudder, dart for the ground. A sudden dart when out about 100 feet from the end of the track ended the flight. Time about 12 seconds (not known exactly as watch was not promptly stopped). The lever for throwing off the engine was broken, and the skid under the rudder cracked."

"As the velocity of the wind was over 35 feet per second," Orville later said, "and the speed of the machine over the ground against this wind ten feet per second, the speed of the machine relative to the air was over 45 feet per second, and the length of the flight was equivalent to a flight of 540 feet made in calm air.

"This flight lasted only 12 seconds, but nevertheless the first in the history of the world in which a machine carrying a man had raised itself by its own power into the air into full flight, had sailed forward without reduction of speed, and had finally landed at a point as high as that from which it had started."

10:42 a.m.

The visitors and the brothers carried the machine back to the track to make repairs for another flight. Because the wind was so cold, everybody was chilled. They all went into the shed to warm themselves around the stove. Johnny Moore looked under the kitchen table and saw a carton filled with eggs. "Where did they ever get so many eggs?" he asked.

Side view of the 1903 machine

"Didn't you notice the little hen running around outside the shed?" asked one of the men from the station. "That chicken lays eight to ten eggs a day!"

Johnny Moore, one of the few people to witness man's first power-driven heavier-than-air flight, rushed outdoors into the cold to look at the wonderful chicken.

11:20 a.m.

After the men repaired the machine, Wilbur made the second trial. "The course was about like mine," Orville wrote in his diary December 17, "up and down but a little longer over the ground though about the same in time. Dist. not measured but about 175 ft. Wind speed not quite so strong. With the aid of the station men present, we picked the machine up and carried it back to the starting ways."

December 17 entry in Orville's diary

10:35 a.m.
December 17, 1903:
the launch of the
twelve-second flight

11:40 a.m.

"I made the third trial," Orville recorded in his diary. "When about the same distance as Will's, I met with a strong gust from the left which raised the left wing and sidled the machine off to the right in a lively manner. I immediately turned the rudder to bring the machine down and then worked the end control. Much to our surprise, on reaching the ground the left wing struck first, showing the lateral control of this machine much more effective than any of our former ones."

12 noon

"Will started on the fourth and last trip," Orville's diary continued. "The machine started off with its ups and downs as it had before, but by the time he had gone over three or four hundred feet he had it under much better control, and was traveling on a fairly even course. It proceeded in this manner till it reached a small hummock out about 800 feet from the starting ways, when it began its pitching again and suddenly darted into the ground. The front rudder frame was badly broken up but the main frame suffered none at all. The distance over the ground was 852 feet in 59 seconds."

After the men removed the front rudder and carried the machine back to camp, they stood discussing the last flight. A sudden gust of wind struck the machine and started to turn it over. Wilbur ran to the front of the machine, but he was too late to stop it. Orville and Daniels seized the uprights at the rear, but the machine turned over on them. Daniels hung on from the inside. Later he said, "I can't tell to save my life how it all happened, but I found myself caught up in them wires and the machine blowing across the beach, heading for the ocean, landing first on one end and then on the other, rolling over and over, and me getting more tangled up in it all the time. I tell you, I was plumb scared. When the thing did stop for half a second I nearly broke up every wire and upright getting out of it.

"The Wright boys ran up to me, pulled my legs and arms, felt on my ribs and told me there were no bones broken. They looked scared."

"His escape was miraculous," Orville said. "The engine legs were all broken off, the chain guides badly bent, a number of uprights and nearly all the rear ends of the ribs were broken."

"I like to think it about now," said John Daniels later. "I like to think about that first airplane the way it sailed off in the air at Kill Devil Hills that morning, as pretty as any bird you ever laid your eyes on. I don't think I ever saw a prettier sight in my life. Its wings and uprights were braced with new and shiny copper piano wires. The sun was shining bright that morning, and the wires just blazed in the sunlight like gold. The machine looked like some big, graceful golden bird sailing off into the wind.

"I think it made us all feel kind of meek and prayerful like. It might have been a circus for some folks, but it wasn't any circus for us who had lived close to those Wright boys during all the months until we were as much wrapped up in the fate of the thing as they were.

"It wasn't luck that made them fly; it was hard work and hard common sense; they put their whole heart and soul and all their energy into an idea and they had the faith. Good Lord, I'm a-wondering what all of us could do if we had faith in our ideas and put all our heart and mind and energy into them like those Wright boys did."

7

Uphill

Does the road wind up-hill all the way?
Yes, to the very end.
— Christina Rossetti, *Up-hill*

Orville Wright to Bishop Milton J. Wright, telegram

Kitty Hawk, December 17, 1903

Success four flights Thursday morning all against twenty-one mile wind started from level with engine power alone average speed through air thirty-one miles fifty-seven seconds inform press home Christmas.

Orvelle Wright

An error in transmission cut two seconds off the longest flight and misspelled Orville.

The two brothers stood around at the weather station while Joseph Dosher sent the telegram through the Norfolk station. As they were about to leave, the telegraph began clicking. The Norfolk operator wanted to know whether he could give the news to a reporter friend. "Absolutely not," said Wilbur. "We want the Dayton papers to have the news first." They left the station, stopped to chat with S. J. Payne at the Life Saving Station, dropped in at the post office, called on a friend who had done some hauling for them, and returned to camp.

At home in Dayton the bishop sat writing at his upstairs desk. It was a little after five and already dark. Downstairs in the kitchen Carrie lit the gas mantle so she could see to start supper. Katharine had gone out Christmas shopping. Someone knocked on the front door. Carrie opened the door, and a messenger handed her a telegram. She hurried upstairs to the bishop's study and handed him the envelope. In a few minutes the bishop came to the kitchen. "Well, Carrie, the boys flew today. They'll be home for Christmas."

"That's nice."

Katharine walked into the kitchen, her pince-nez lenses steaming. "It's nippy out." The bishop held out the telegram. "What's this?" She put down her packages.

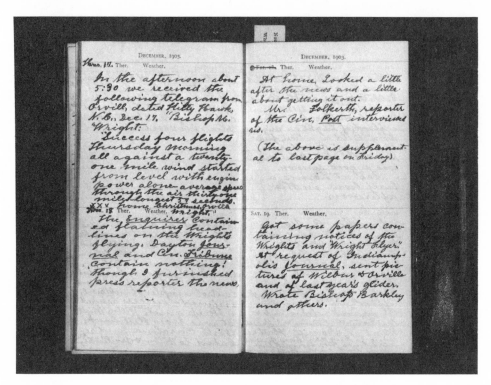

*December 17, 18, 19
entries from 75-year-old
Bishop Wright's diary,
now in the Wright
Brothers Collection,
Wright State University,
Dayton*

After she read the telegram, Katharine turned to Carrie. "I think you'd better hold up supper for a while. Papa, we must take this over to Lorin's and I must also send a telegram to Mr. Chanute."

When the bishop and Katharine arrived at Lorin's home, the family was at the table. Lorin read the telegram. "I'll take it down to the *Journal* office after supper," he told them. Turning to the children, he said, "Uncle Orv and Uncle Will flew today and will be home for Christmas."

The general public, including newsmen, were not ready to accept the news that two men at Kitty Hawk had actually flown in a heavier-than-air powered machine. Simon Newcomb, the world's foremost astronomer of the day, had stated in print that flying was nonsense and simply couldn't be done. When Langley's attempt had failed, Newcomb again stated that flight was impossible and explained why. People believed him.

Lorin Wright walked to the *Journal* office and asked for Frank Tunison, who looked at the telegram and said, "Fifty-seven seconds? If it had been fifty-seven minutes, it might have been a story." He shook his head and tossed the telegram on the desk.

Lorin walked back home. There was no story in the next morning's *Journal.*

The telegraph operator in Norfolk, however, saw a story in the telegram. Ignoring Wilbur's "Absolutely not," he called H. P. Moore, a circulation manager for the *Virginian-Pilot,* who tried without success to get information about the flight from sources at Kitty Hawk. He then fabricated his own version, describing a flight from a platform on a high sand hill with Wilbur piloting the craft to a height of 60 feet. The story, rewritten by reporter Ed Dean, appeared in the *Pilot* the next morning under the banner headline: Flying Machine Soars 3 Miles In Teeth Of High Wind Over Sand Hills And Waves At Kitty Hawk On Carolina Coast, No Balloon Attached To Aid It.

Moore offered the story to twenty-one papers in the country. Five ordered it and of those five, only the *Cincinnati Enquirer* printed it. The *Dayton Evening Herald* carried on its front page the headlines:

In honor of the 75th anniversary of powered flight this souvenir was produced in 1978.

DAYTON BOYS
FLY AIRSHIP

Machine Makes High Speed in the
Teeth of a Gale and Lands at
the Point Selected

PROBLEM OF AERIAL NAVIGATION SOLVED
Ascent Made at Isolated Spot on Carolina Coast, Where Wrights Experimented for Three Years

The story said, "The mile was covered and then Orville Wright declared the invention a success, but it was not until a third (mile) had been accomplished that the inventor

cast his eyes about for a suitable landing, found it and with his invention under utmost control, slowly neared the earth and let his machine light as easily and gracefully as a bird.''

The *Dayton Daily News* had no story on December 18 but on the following day a story on page one said:

THE WRIGHT BOYS
ARE COMING HOME

Norfolk, Va., Dec. 19—Orville and Wilbur Wright, inventors of the Wright "Flyer," which made several successful flights near here Thursday, left today for their home in Dayton, O., to spend Christmas with their parents.

While the brothers were packing on December 18, Wilbur and Orville received telegrams from the *New York World,* the *Woman's Home Companion, Scientific American,* and *Century Magazine,* all wanting photographs of the flights.

Octave Chanute to Wright Brothers (telegram)

Chicago, December 18, 1903

Immensely pleased at your success. When ready to make it public please advise me.

The brothers arrived home Wednesday night, December 23. The family spent Christmas Day at Lorin's home, and Wilbur carved the turkey. Two days later Octave Chanute sent an invitation to the brothers to give the first scientific report on the flights at a convention of the American Association for the Advancement of Science at its winter meeting in St. Louis December 28 to January 2.

Wilbur replied by telegram that they were giving no pictures nor descriptions of the machine or their methods at that time. He also sent Chanute a long letter for his information alone, describing the events from the time when Chanute had left Kitty Hawk on November 12, 1903, through December 17.

Wright Brothers to the Associated Press

January 5, 1904

It had not been our intention to make any detailed public statement concerning the private trials of our power Flyer on the 17th of December last; but since the contents of a private telegram, announcing to our folks at home the success of our trials, was dishonestly communicated to the newspaper men at the Norfolk office, and led to the imposition upon the public by persons who never saw the Flyer or its flights, of a fictitious story incorrect in almost every detail; and since this story together with several pretended interviews or statements, which were fakes pure and simple, have been widely disseminated, we feel impelled to make some correction.

Wilbur then gave the Associated Press a detailed account beginning with the events of the morning of December 17, going through the four flights and concluding thus:

When these points had been definitely established, we at once packed our goods and returned home, knowing that the age of the flying machine had come at last.

From the beginning we have employed new principles of control; and as all the experiments have been conducted at our own expense without assistance from any individual or institution, we do not feel ready at present to give out any pictures or detailed description of the machine.

The tone of Wilbur's letter and his attitude annoyed the Associated Press to the extent that when it distributed Wilbur's account, the Press omitted the first paragraph. The brothers' unwillingness to provide photographs when they were first asked contributed to the cold reception given to the Wright brothers by members of the press for many years.

Not only did Wilbur's letter anger the Associated Press but the Wright brothers themselves were angered by a letter they received. Augustus Herring wrote, claiming that he not only originated the glider but also he had independently reached the solution of powered flight and because of this he offered in his letter to share a three-way partnership with the Wrights. "This time he surprised us," Wilbur wrote to Chanute. "Before he left camp in 1902 we foresaw and predicted the object of his visit to Washington, we also felt certain that he was making a frenzied attempt to mount a motor on a copy of our 1902 glider and thus anticipate us, even before you told us of it last fall. But that he would have the effrontery to write us such a letter, after his other schemes of rascality had failed, was really a little more than we expected. We shall make no answer at all."

"I am amazed at the impudence of Mr. Herring," Chanute answered on January 14, 1904, "in asking for one-third of *your* invention. While I could wish that you had applied for the patents when I first urged you to do so, I think that your interests are quite safe. The fact that Mr. Herring visited your camp, in consequence of circumstances which I subsequently regretted, will certainly upset any claims which he may bring forth. I suppose that you can do nothing until an interference is declared. If it is, please call on me, and in the meantime try to find out who is his patent attorney.

"In the clipping which you sent me: 'All the experiments have been conducted at our own expense without assistance from *any* individual or institution.'—Please write me just what you had in your mind concerning myself when you framed that sentence in that way."

Sensing that Chanute's feelings had been hurt by the statement, Wilbur hurried to explain to him that what he meant was since they had paid for their own experiments, they were entitled to refuse to make their discoveries public property in contrast to Langley, who because his work had been done with a government grant, could not refuse to publish his discoveries. "The use of the word 'any,' which you underscored, grew out of the fact that we found from articles in both foreign and American papers, and even in correspondence, that there was a somewhat general impression that our Kitty Hawk experiments had not been carried on at our own expense. We thought that it might save embarrassment to correct this promptly."

Early in January 1904 the brothers began drawing plans for a new machine which they expected to use in demonstrations and further experiments. They wrote for rules for the flying contest at the St. Louis Exposition. Late in February they went to St. Louis to inspect the site. They found that the course was laid out so that if a forced landing occurred, serious damage to the machine would be inevitable.

If a man-carrying machine did not win the first prize, furthermore, it could win none at all, whereas subsidiary prizes were offered for dirigible balloons, kites, gliders, engines and even toy planes. "A flight of even one mile by a man-carrying machine would be an event of great importance in aeronautical history and yet the rules would give it no recognition even to the extent of a brass medal," Wilbur told the committee. When they returned home, they had not decided whether they would enter.

The February 4 issue of *Independent* carried an article, "Experiment of a Flying Man," signed by Wilbur Wright. Wilbur, who had not written it, wrote to D.A. Willey, saying, "I never in my life wrote you a letter; I never in my life gave you authority to use my name in any way whatever; I never gave you authority to make extended verbatim quotations from my addresses before the Western Society of Engineers, which are copy-righted by that Society. Neither has the Society given you such permission, the secretary tells me.

"The *Independent* will probably furnish a statement regarding the matter. If not, I will. Meantime you can take whatever course you think will profit you best."

The magazine writer decided to publish a retraction. Chanute admitted to Wilbur that he had furnished Willey with some information but he himself had been duped, not knowing that the story was to be published as by Wilbur. Chanute suggested that Wilbur consult a lawyer.

But Wilbur had to go to Huntington three days in May in connection with the church trouble. In addition, his time was taken in trying to locate a field near Dayton where he and Orville could fly their new machines. They found one at Simms Station about eight miles outside the city limits. It was a cow pasture belonging to Torrence Huffman, who said they might use it provided they did not run over his cows.

So far the brothers had made no money from their flying machine, but now they began to anticipate possibilities. The French Aero-Club announced in April a grand aviation prize of $100,000. While Chanute thought it might be limited to French aviators, he offered to find out for the Wrights. They indicated they were interested even if they had to go to France to compete.

The 1904 machine was ready for testing in May. The dimensions were similar to the 1903 machine, the weight was increased to 900 pounds, a new 18-horsepower engine was added, and the shape of the vertical rudder was changed. When the brothers tested the machine in June they moved the engine, gas tank, and radiator in order to change the center of gravity towards the rear.

When the 1904 machine was ready for its first test, the brothers sent letters to the Dayton and Cincinnati papers, saying that they would be glad to have press representa-tives watch their flight on May 23. Twelve reporters came to the field along with several others the brothers had invited, including the bishop.

Orville and Wilbur stand beside their 1904 machine at Huffman Prairie, Dayton.

The 1904 Flyer on the launching track at Huffman Prairie [Simms Station], summer 1904

The trial failed. At first the wind velocity was too high and then it was too low. They gave up the attempt for the day. A few reporters returned the next day for the second attempt. This time the machine rose five or six feet from the ground and flew for nearly sixty feet before it came down.

Wilbur to Chanute

Dayton, June 21, 1904

At Kitty Hawk we had unlimited space and wind enough to make starting easy with a short track. If the wind was very light we could utilize the hills if necessary in getting the initial velocity. Here we must depend on a long track, and light winds or even dead calms. We are in a large meadow of about 100 acres. It is skirted on the west and north by trees. This not only shuts off the wind somewhat but also probably gives a slight downtrend. The greater troubles are the facts that in addition to cattle there have been a dozen or more horses in the pasture and as it is surrounded by barbwire fencing we have been at much trouble to get them safely away before making trials. Also the ground is an old swamp and is filled with grassy hummocks some six inches high so that it resembles a praire-dog town. This makes the tracklaying slow work. While we are getting ready the favorable opportunities slip away, and we are usually up against a rainstorm, a dead calm or a wind blowing at right angles to the track. Today we had our first decent chance, but as the margin was very small we were not skillful enough to get really started. The first two flights were for a distance for a little more than a hundred feet and the third two hundred and twenty-five feet. On this one Orville almost got away, but after 200 ft., he allowed the machine to turn up a little too much and it stalled it. He had a speed of about 18 miles on leaving the track, but the rise necessary to gain a little room for maneuvering reduced this to about 16 miles and as the wind was blowing only 8 miles, and unsteady at that, the resistance was too high to permit rapid acceleration, owing to the great angle of incidence required. To get started under such conditions requires perfect management. We are a little rusty. With a little more track and a little more practice we hope to get a real start before long and then we will see what the machine can really do in the way of flying.

We have about concluded to enter the St. Louis contest but are reluctant to do this formally, until we are certain of being ready in time. We have one machine finished, another approaching completion, and a third well started. As these are built to measure, the parts are interchangeable, and even a rather serious accident would not necessarily throw us out of the contest. If the Exposition people will hold the door open till we get ready, there is yet hope that there may be a real contest for the grand prize.

Because of the light winds at Simms Station the brothers realized that they would have to build a starting device that would make them independent of wind. Early in September they erected a derrick that supported a 1,600-pound weight. When the weight dropped, it put a 350-pound pull on the engine, enabling it to take off in 50 feet.

Once they began using the derrick, their flights became more successful each day. On September 20 Orville flew the first complete circle, a distance of 4,080 feet. This flight was witnessed by Amos I. Root, seller of beekeeper supplies and honey at Medina, Ohio. In *Gleanings in Bee Culture,* a magazine he published, he had written an account of the first flight at Kitty Hawk. In September he drove from Medina to Simms Station, saw Orville fly the circle and stayed over for several days to witness more flights. The January 1, 1905, issue of his magazine published the first eyewitness account of any successful powered flight. He sent a copy to the *Scientific American* with permission to reprint. The editor ignored Root's account.

Wilbur, reporting to Chanute the success circumnavigating Huffman Prairie, said that the news that they were flying every day was getting about in the neighborhood. Since the brothers had not yet been awarded a patent, they wanted to keep their flying secret for the time being, and they were becoming uneasy about continuing at Simms Station. "In fact," said Wilbur, "it is a question whether we are not ready to begin considering what we will do with our baby now that we have it."

Early in January 1905 Wilbur spoke to Congressman Robert M. Nevin of Dayton in an attempt to interest the government in the Wright Flyer. Nevin suggested that Wilbur write a letter which Nevin would give to President William Howard Taft. In his letter Wilbur pointed out that flying had advanced to a point where it could be used in scouting and carrying messages in war time. He offered to furnish the government machines at a contract price or to furnish all the information he and his brother had acquired, together with a license to use the Wright patents whenever they were granted.

The 1904 machine flying
at Huffman Prairie
November 16, 1904

Nevin wrote to the Wrights January 26, enclosing a letter from Major General George L. Gillespie, United States Board of Ordnance and Fortifications, which said:

I have the honor to inform you that, as many requests have been made for financial assistance in the development of designs for flying-machines, the board has found it necessary to decline to make allotments for the experimental development of devices for mechanical flight, and has determined that, before suggestions with that object in view will be considered, the device must have been brought to the stage of practical operation without expense to the United States.

It appears that from the letter of Messrs. Wilbur and Orville Wright that their machine has not yet been brought to the stage of practical operations, but as soon as it shall have been perfected, this Board would be pleased to receive further representations from them in regard to it.

Since the United States government showed no interest in the Flyer and was in fact ignorant of the Wrights' successful flights in the past two years, the brothers on March 1, 1905, offered to furnish a scouting plane to the British War Office. Two months later the British sent word that the military attache in Washington would call on them to witness a flight.

Wilbur Wright to Octave Chanute

June 1, 1905

It is no pleasant thought to us that any foreign country should take from America any share of the glory of having conquered the flying problem, but we feel we have done our full share toward making this an American invention, and if it is sent abroad for further development the responsibility does not rest upon us. We have taken pains to see that Opportunity gave a good clear knock on the War Department door. It has for years been our business practice to sell to those who wished to buy, instead of trying to force goods upon people who did not want them. If the American government has decided to spend no more money on flying machines till their practical use has been demonstrated in actual service abroad, we are sorry, but we cannot reasonably object. They are the judges.

The brothers completed the 1905 machine in June. Its design was similar to the 1903 and 1904 machines with enlarged horizontal controls and rudder and a stronger framework. They tried three different kinds of propellers, finally adopting tips, which they called "Little Jokers" and later "blinkers." These were small surfaces set at an angle to balance the pressures that distorted the blades, causing sideslip on turns.

Orville made a series of forty-nine flights with the 1905 machine, several of which resulted in slight damage to the machine on landing. On the July 14 flight, the machine suddenly turned downward, breaking the front skids, front rudder, upper front spar, one upright and a number of ribs. Orville flew through the broken top surface but got to his feet, jolted but uninjured.

France had been enthusiastic over aviation ever since Chanute spoke to the Aéro-Club de France in 1903. Members of the club, led by Ernest Archdeacon, had been trying to be first in the world to fly. In the spring of 1905 Archdeacon wrote to the Wrights that he did not believe they had flown and said that he would willingly come to America to see them fly. He mentioned that he and Henri Deutsch de la Meurthe had established a 50,000 franc prize for the first flight and challenged the brothers to allow themselves to be seen in America or to come prove themselves in France. Chanute was amused by the letter and said that he had an idea that if the Wrights invited one or two judges to come to America to see them try for the prize, they would not accept.

1905 Machine

The 1905 machine had a wingspan of 40 feet, 6 inches, a wing area of 503 square feet, horizontal rudder area of 83 square feet and vertical rudder area of 34.8 square feet. It was 28 feet long and weighed 710 pounds.

At Kitty Hawk in 1908 it was altered so that the operator and a passenger could sit upright on the lower wing surface.

By October 1905 the brothers had made more than fifty successful flights from Simms Station. On October 4 Orville flew almost twenty-one miles in thirty-three minutes, twenty seconds. Wilbur made the longest flight of the year, 29 rounds of the field, 24½ miles in 39 minutes, 24 4/5 seconds. Four days later—October 9—they wrote to the Secretary of War, saying that they did not wish to take their invention abroad unless they found it necessary to do so, and therefore they were writing again to renew their offer to furnish machines on contract to be accepted only after trial trips. The machine would carry an operator and supplies of fuel sufficient for one hundred miles. They were also willing to take contracts building machines carrying more than one man.

When the reply came from the Board of Ordnance it was the same old answer. It was identical in wording with the first paragraph of the letter they had earlier received, repeating that the United States did not make grants to any one to develop a flying machine. The Board also suggested that the Wrights would have to furnish them the approximate cost of the complete flying machine, the delivery date and provide drawings and descriptions to enable the Board to conclude whether the machine was practicable.

Wilbur Wright's summary of the 1905 experiments describes trouble with stalling of the machine when it flew in circles. "The remedy was found to consist in the more skillful operation of the machine and not in a different construction. When we discovered the real nature of the trouble, and knew that it could always be remedied by tilting the machine forward a little, so that its flying speed would be restored, we felt that we were ready to place flying machines on the market."

Flight 41, a 12-mile flight,
Huffman Prairie,
September 29, 1905

Flight 46: an over-heated bearing ended Wilbur's 24½-mile flight at Huffman Prairie. It was the longest flight of 1905.

Wilbur Wright to Octave Chanute

Dayton, December 8, 1905

Word just received from our attorneys is to the effect that all our claims have been allowed in U.S. patent, and that the case will be ready for issue as soon as a few unimportant corrections have been made in the wording.

On further consideration we are inclined to think that the best course regarding the correspondence with the American War Department is complete suppression. We do not think there would be any advantage in bringing the matter to the attention of the President, or Sec. of War, unless they were previously *fully convinced* of the practicability of the machine and the desirability of securing it for the government. No such condition exists at present. Neither can we see any advantage in letting it become generally known that we have been turned down by our government. It will be a hindrance to successful negotiations with any other government. We have not informed a single person outside of our own family, except yourself. We think it best to maintain secrecy as to the progress of all negotiations with governments.

Negotiations with the French government began three days after Christmas 1905 when Arnold Fordyce, representing a French syndicate, arrived in Dayton. He explained to the Wrights that the syndicate planned to present a Wright Flyer to the French government as a gift for war purposes. The Wrights agreed to deliver a Flyer no later than August 1, 1906, for $200,000 and to demonstrate it in France, stipulating that they would give a license to manufacture machines to the government only.

Wright Flyer III [1905] in process of restoration for exhibit at Carillon Park, Dayton

As soon as news of the agreement with the French syndicate became public, the Austrian Association of Builders asked to purchase a Flyer as a gift for Emperor Franz Josef on his sixtieth Jubilee in 1908. Wilbur wrote to Chanute, "If the idea of acquiring the machine by different countries by popular subscription should spread, we may be able to secure all the remuneration we care for, and establish free trade within a year or two. Nothing would suit us better, but we shall not begin counting our chickens until we are sure of them."

Stories of the Wrights' accomplishments, some true, some false, appeared in newspapers and periodicals all over the world. Every day the mailman brought letters by the score to 7 Hawthorn Street. Orville wrote to Frank S. Lahm, member of the Aéro-Club de France, that they had read in a French clipping that when Ernest Archdeacon learned about the syndicate's offer to the brothers, he made an offer to save the French government 800,000 francs by having 200,000 francs appropriated to his own experiments. Archdeacon stated that he and Captain Ferdinand Ferber, who had earlier tried to buy a Flyer, could build a flying machine in three months. "For a man who has just passed through four years of failure," said Orville, "he seems rather sanguine." The Wrights were aware that Ferber, Archdeacon and other French aviation enthusiasts did not have the experience to build a practicable machine.

Aviation enthusiasts all over the world were saddened when Samuel P. Langley died February 27, 1906, in his seventy-second year, never having seen one of his machines fly. "No doubt disappointment shortened his life," Wilbur said. "It is really pathetic that he should have missed the honor he cared for above all others, merely because he could not launch his machine successfully. If he could only have started it, the chances are it would have flown sufficiently to have secured to him the name he coveted, even though a complete wreck attended the landing. I cannot help feeling sorry for him. The fact that the great scientist, Prof. Langley, believed in flying machines was one thing that encouraged us to begin our studies.

"He possessed mental and moral qualities of the kind that influence history. When scientists in general considered it discreditable to work in the field of aeronautics he possessed both the discernment to discover possibilities there and the moral courage to subject himself to the ridicule of the public and the apologies of his friends. He deserves more credit for this than he has yet received."

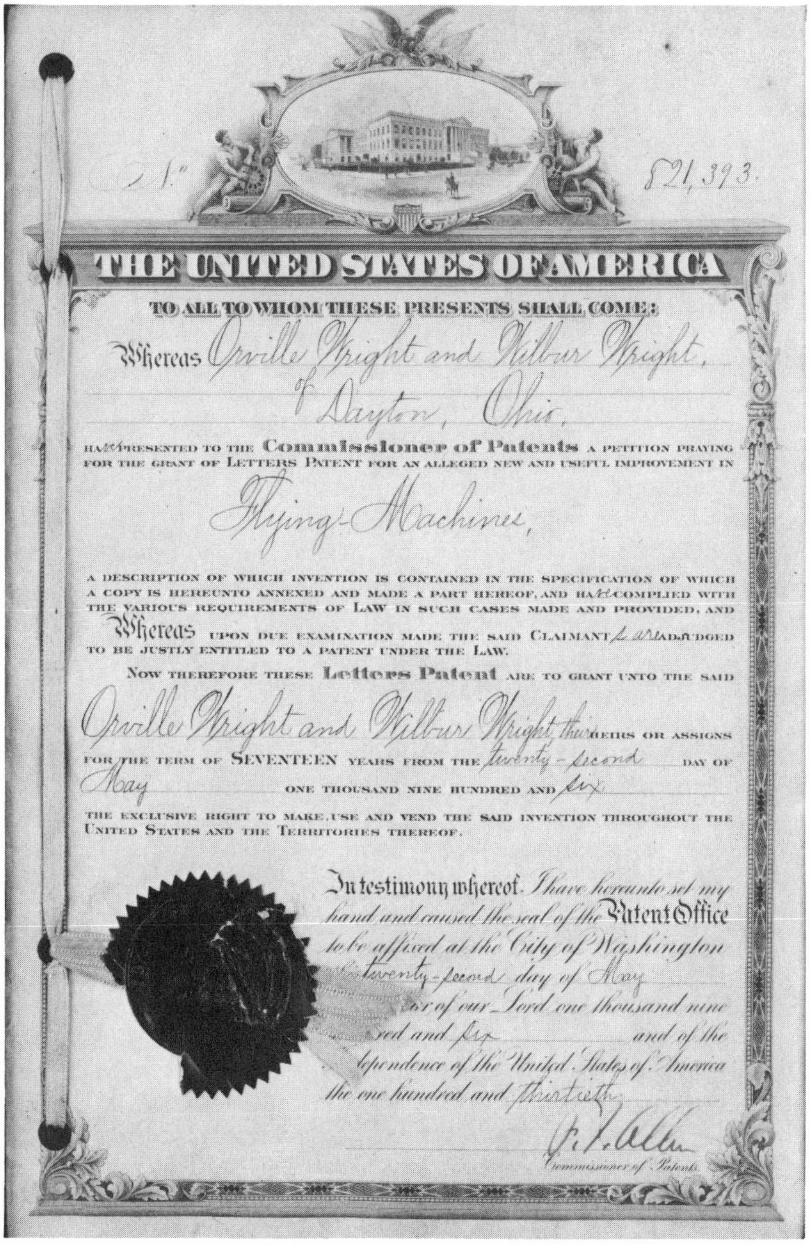

*U.S. Patent 821, 393
granted to the Wrights,
May 22, 1906*

Men from the French government arrived in Dayton in April to discuss amendments to the contract they had made with the Wrights. Because the threat of a war between France and Germany over Morocco had lessened, the French wished to cancel the contract. They made such unreasonable demands of performance that the Wrights could not agree, and the contract lapsed by default although the brothers collected $5,000 penalty.

Patrick Y. Alexander, an agent for the British government, made a trip to Dayton in April and was a dinner guest at the Wright home. The brothers suspected he had come to find out whether they had agreed to sell to the French.

The United States Patent Office granted the Wrights patent 821,393 for a flying machine May 22, 1906. They had filed the application March 23, 1903. The government of Germany issued patent 22,051 on July 16, 1906.

Glenn H. Curtiss, twenty-eight years old in 1906, owned the G. H. Curtiss Manufacturing Company in Hammondsport, New York, which made engines for bicycles, motorcycles, and dirigibles.

Glenn H. Curtiss to the Wright Brothers

May 16, 1906

We have built a large number of motors for aerial purposes. We have sold motors to Stevens, Knabenschue, Tomlinson and others; while Baldwin has used ours exclusively for the last two years in 23 successful (dirigible) flights at Portland, Ore.

We understand that your work is of a somewhat different character, but we mention these to prove that our motor has great power and reliability, as it has proved that much power is required to drag a big gas bag through the air at 15 mph.

We recently shipped to Dr. Bell's Nova Scotia laboratory a motor designed for airplane work, and we hope this experience will be of service to you.

The writer will be in Columbus next Sunday, and if you are interested would be willing to go to Dayton and take up the matter further with you. Address me, care of Oscar Lear Auto Co.

Glenn H. Curtiss
1867—1930

The Wrights mailed their reply to the Hammondsport address given on the company letterhead. Finding no answer waiting in Columbus, Curtiss telegraphed the Wrights: If convenient like to talk with you 6 o'clock Bell phone.

That evening Curtiss telephoned the Wrights. One of the brothers said they were not interested in the engine but would welcome him in Dayton. Three months later, in September 1906, Curtiss accompanied Thomas Baldwin to Dayton to service the Curtiss engine on Baldwin's dirigible. Baldwin had a week's contract at the Montgomery County Fair, where Wilbur and Orville saw the balloon in flight.

Baldwin brought Curtiss to the Wright Cycle Shop and the four men held a long chat. Curtiss asked questions about propellers, and the Wrights freely told him how to improve their performance by cutting away part of the inner surfaces of the blades in order to reduce air resistance.

After Curtiss left Dayton he wrote to the Wrights on September 11 that he had used their suggestions and the propellers on Baldwin's dirigible had been greatly improved.

In years to come, the name of Glenn H. Curtiss would be a bugbear to the Wright brothers.

8

New Clothes for the Emperor

I cannot hobnob with the Emperor
when I go to Berlin without some clothes.
                              —Wilbur Wright, 1907

"Are the Wright brothers honest? Are they dependable?"

To find out, New York banker Ulysses Eddy visited them in Dayton at Thanksgiving 1906.

After talking with the brothers for a time, he made up his mind. "Sirs," he said, "the company headed by Charles R. Flint might work with you in developing your business and if you are interested, I will speak with Mr. Flint in your behalf."

"Who is Mr. Flint?" Wilbur asked. "What does he do?"

"He heads a company of bankers, munitions makers and promoters of new products."

The brothers agreed to meet with Mr. Flint.

In response to a telegram from Flint, Orville left for New York December 17.

Wilbur to Chanute

                              Dayton, December 20, 1906

Flint & Co. now offer us $500,000 for all rights outside of the United States. We reserve the latter. The money to be paid in cash upon the delivery of one machine after a demonstration consisting of a flight of 50 kilometers. We are to have the privilege of going for prizes, etc., and to publish anything we choose after a limited time. Their idea seems to be to depend on getting possession of the market by being first in the field rather than by depending on patents alone or secrecy alone. Orville has just returned from New York where he had a conference with them this week. We have not committed ourselves. What do you think of it? Do you think them safe people to deal with under proper precautions? The price and terms are satisfactory and we would accept if we felt sure of their character.

Chanute answered that he had known Flint for twenty-five years, but not well. He thought the terms offered were better than he expected, but he wondered whether Flint might allow the invention to pass into the hands of a single nation and if it did, perhaps a moral question might be involved. He had learned that perhaps Flint would negotiate with Russia and feared that Russia's purchase of the flying machine might "mean a new war against Japan and much bloodshed."

The brothers went to New York in January for further talks with Flint, who proposed at that time that if they would give a private exhibition for the Czar of Russia, he would guarantee them $50,000 and 90 percent of any orders up to $500,000 and 50 percent of any orders after that. Flint himself would conduct the business with Russia, and in Europe he planned to work through other agents. They also discussed an alternate plan whereby the Wrights would retain control of the terms and places of selling and the Flint company would work under their direction.

Chanute thought that Flint's offer for the sales to Russia was a good one, but cautioned that their idea of otherwise retaining full control of the terms and places of selling was unwise. "You can lay down the general terms required in your agreement with him and you can enjoin being consulted about deviations asked for," he said, "but the agent must have some latitude to effect sales."

Wilbur went alone to New York in February 1907. Flint called off the demonstration and negotiations with Russia to await a decision on a proposal he had made to the German government to furnish 50 machines at $50,000. "I will probably stay here another day," he wrote Orville, "so that we may consider what country to try next in case there is no encouragement from Germany." He returned home without any proposals to consider.

A letter from Herbert Parsons, a Republican Congressman from New York, asked to see copies of the correspondence between the Wrights and the Board of Ordnance in 1905. When he sent it, Wilbur wrote, "We are prepared to give courteous consideration to any request the Board may make of us, but its action of October 24, 1905, renders it impossible for us to make any request of it." The reference was to the letter they had received from the Board saying that it did not care to formulate any requirements for the performance of a flying machine or take any further action on the subject until a machine was produced "which by actual operation is shown to be able to produce horizontal flight and carry an operator." On October 5, 1905, Orville had flown thirty rounds of the field at Simms Station in the presence of about 20 witnesses. The bishop had written in his diary that he saw Orville fly twenty-four miles in thirty-eight minutes, one second. The obtuseness of the Board of Ordnance so angered the brothers that they agreed they would never again initiate any proposals to the United States government.

They learned that President Theodore Roosevelt and many Army and Navy officers were scheduled to observe a naval review at Hampton Roads April 26, 1907, as part of a celebration of the three-hundredth anniversary of the founding of the first English colony at Jamestown.

"Why don't we assemble a new machine at our old camp at Kitty Hawk," said the mischievous brother Orville, "fly it from there to Jamestown and after we take an unexpected part in the parade, fly it back to the camp. Nobody will know where we came from, who we are or where we went."

"To do that we'd have to build a hydroplane," said Wilbur. "We can land on Currituck Sound."

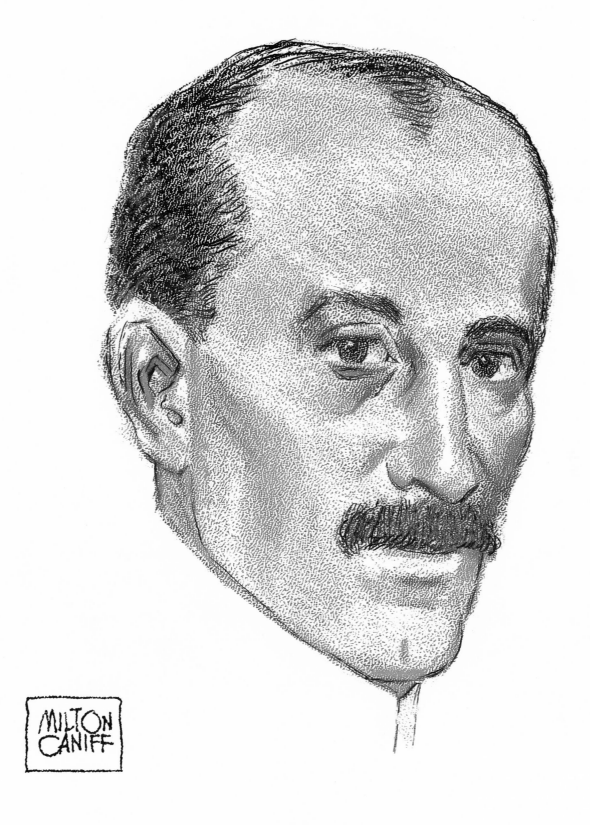

Orville Wright

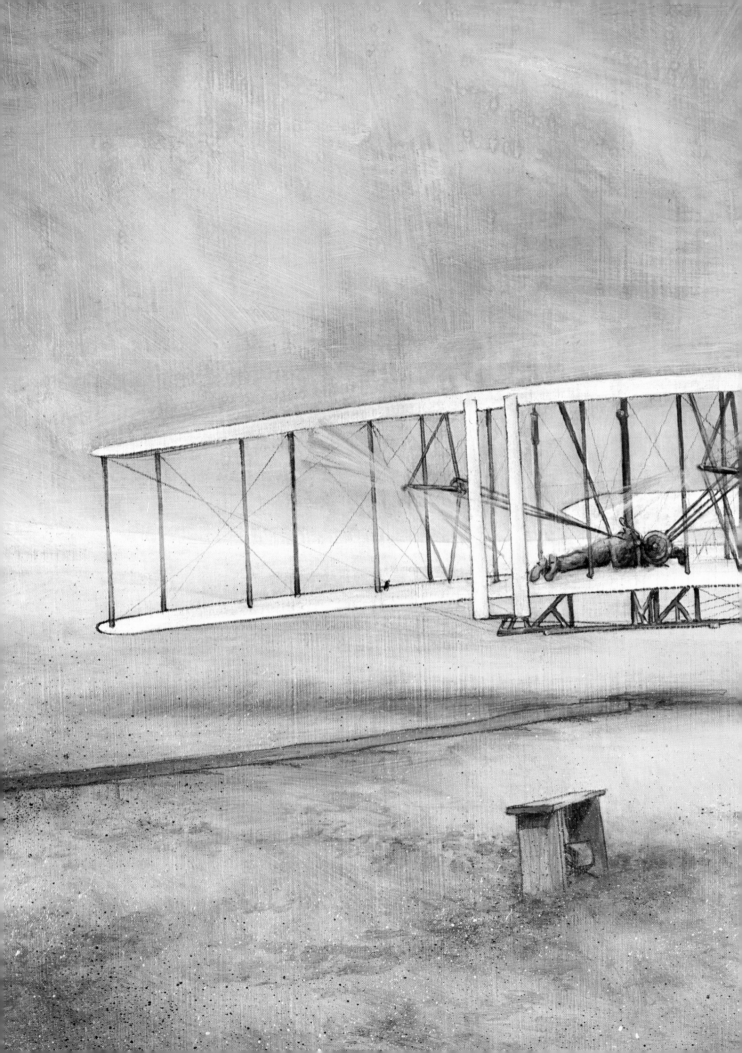

FRANK PAUER

Wilbur Wright

They fitted a machine with pontoons and experimented taking off from the Miami River in Dayton. Crowds of spectators gathered on the Third Street bridge and lined the river banks. When the brothers attempted to take off, the bottom of the floats dragged on the water. As the front edges reached the surface of the water, the floats sank because of a loss of lift on the upper side when the water ceased flowing over the top. One of the propellers broke, and the brothers gave up the idea.

Bishop Wright's Diary

May 16, 1907

I was at home. Wilbur got a telegram from Flint, and at 10 p.m. he started for New York to take a ship for London. He goes to talk with agents, in London, Paris and Berlin.

En route to New York Wilbur wrote to Chanute that before he left home they had received a communication from the Board of Ordnance "notifying us that the Board had under consideration several propositions for the construction and testing of flying machines, and that we could take any action in the matter we chose. We are not sanguine that the Board is really in a mood to reach an agreement, but we will give the matter serious attention.

"It is reported by Dr. Bell that the Langley machine is to be tried again, but he does not say whether the War Department is to have a hand in it."

Orville Wright to the Board of Ordance and Fortification, War Department

Dayton, May 17, 1907

We have some flyers in the course of construction and would be pleased to sell one or more of them to the War Department, if an agreement as to terms can be reached.

These machines will carry two men, an operator and an observer, and a sufficient supply of fuel for a flight of two hundred kilometers. We are willing to make it a condition of a contract that the machine must make a trial trip before Government representatives of not less than fifty kilometers at a speed of not less than fifty kilometers an hour, before its acceptance by the Department, and before any part of the purchase price is paid to us.

If the War Department is in a position to purchase at this time, we will be pleased to have a conference for the purpose of discussing the matter in detail, or we are willing to submit a formal proposition, if that is preferred.

Parisian Hart O. Berg had conducted a number of negotiations for Charles Flint and Company, and it was he Flint hoped would be able to promote the sale of Wright Flyers in Europe. But Berg was not much interested because like so many others, he had doubts that the Wrights had actually flown. Flint believed that if Berg met even one of the brothers, he would change his mind. Berg agreed to do so.

Wilbur arrived in Liverpool May 25 and took the train to London. Berg was waiting for him at Euston Station. Although Berg had never seen a picture of Wilbur, he picked him out of the crowd immediately. "Either I am a Sherlock Holmes," Berg wrote to Flint, "or Wright has that peculiar glint of genius in his eye which left no doubt in my mind as to who he was."

Berg took Wilbur to Morley's Hotel opposite the Nelson Monument in Trafalgar Square. Wilbur had arrived with only one suitcase—Berg described it as a music roll—and realized that he had not brought along enough clothes. Berg took him to a tailor in the Strand. "I ordered a dress suit today for 7 pounds and a dinner jacket for 4 pounds," he wrote Orville. "They are the best the tailor had and I think would cost close to a hundred dollars at home." Orville wrote later, "I would give three cents to see you in your dress suit and plug hat." Three cents was as much as Orville ever gambled and then it was only figurative.

After the shopping Berg and Wilbur spent the evening talking. Wilbur's reluctance to commit himself puzzled Berg. Both men felt that there was little to be hoped for at the time from England or France. "There was only the German government left," Berg told Flint, "and even there I assured him that the government would do nothing, but we must look to the power greater than the government, that is, the Emperor himself. I proceeded to explain to him that if the Emperor did recommend a full examination of his apparatus I was fearful that the Aeronautic Officers in the German Army would be apt to put all sorts of difficulties in the way, and I was fearful that it would be a long-winded affair.

"About 5 o'clock in the afternoon, I think, you will distinctly note that I say 'I think,' I brought some sort of action in his mind and think he was on the point, you will note I distinctly said 'I think he was on the point,' of veering around from the company to government methods."

Berg told Wilbur he had made an engagement for him to talk with Deutsch de la Meurthe, who had asked that Wilbur should see him before anyone else. "I think he agreed, you will note I distinctly say 'I think he agreed,' to go to Paris with me Monday. I am to see him at 1 o'clock today, Sunday, and I think I shall be able, you will kindly note that I distinctly say that 'I think I shall be able,' to get a more distinct expression from him of what he wants than resulted in my efforts of yesterday."

Berg disagreed with Wilbur that if they offered to sell to private individuals and were refused by them their chances with governments would be diminished. Berg thought that they should not go so far with governments as to get a refusal from them because then their chances of organizing a sales company would be poor since they could not offer the possibility that the company could get government business. "The idea being refused by private individuals would have no influence on governments, but being refused by governments would have a great influence on private individuals. I think Mr. Wright eventually looked at it in this light—you will note that I distinctly say that 'I think Mr. Wright, etc.'" Berg added in his letter to Flint that he liked Wilbur's personality and thought he inspired great confidence and would be a capital Exhibit A.

Wilbur and Berg left the next day for Paris. In letters to Orville, his father and Katharine, Wilbur described his sightseeing as precisely as he wrote about their flying experiments.

"I spent a couple hours in the Louvre," he wrote. "I liked the Rembrandts, van Dycks, Holbeins, Dous, Jordaens—a whole lot better than the Reubens, Titians, Raphàels and Murillos. There seems to be a great difference between different pictures by the same man. The Mona Lisa is no better than the prints in black and white. I must confess that the pictures by celebrated masters that impressed me most were not the ones that are best known. I like da Vinci's *John the Baptist* much better than his *Mona Lisa*. Van Dyck is really the best portrait painter as a whole, though his *Children of Charles I* is not near as good as the prints, and his celebrated portrait of Charles is inferior to some of his other paintings.

"The French understand how to place public buildings. There is always an open space. And in addition there is nearly always a broad avenue leading directly up to it, giving a view from a long distance. In Paris the parks are the play grounds of the people.

In day time they are filled with children wearing little black silesia aprons or rather coats coming about to their knees, to keep their clothes clean. It is very amusing, but shows French thrift.

"I also visited rooms containing the French paintings of the 19th century. While I do not pretend to be much of a judge I am inclined to think that in five hundred years it will be recognized as some of the greatest work ever done. Corot and Millet have styles peculiar to themselves. Corot nearly always uses a very light background of sky with the greatest light close to the ground. His trees are hazy and stand between you and the light, so that the latter comes through the trees instead of falling on them. The pictures of Millet are usually figures of peasants or workmen not very clearly outlined on a dark background. The *Angelus* and the *Gleaners* are really not in his most common style.

"I stopped to have a look at the inside of the Notre Dame. It was rather disappointing as most sights are to me. My imagination pictures things more vividly than my eyes. The nave is seemingly not much wider than a store room and the windows of the clerestory are so awfully high up that the building is very dark. The pillars are so heavy and close together that the double aisles on each side form no part of the room when you stand in the nave. The latter is so high that the length seems shorter than it really must be."

When Wilbur and Henri Deutsch de la Meurthe met in Paris, Deutsch said he was ready to join in the formation of a company or else to take Wilbur to the Minister of War to determine whether the government was interested. But Wilbur felt that since he had already talked with Arnold Fordyce, who had come to Dayton to see the brothers in 1905, he should see Fordyce first. Fordyce was associated with Henri Letellier, owner of *Le Journal*. Deutsch owned a rival newspaper, *Le Matin*. When Letellier learned that Wilbur had talked to Deutsch, he refused to cooperate in any company that included Deutsch. Deutsch called on the Minister of War, General Georges Picquart, who said he was interested in buying a machine if the Wrights could guarantee it would fly at a height of three hundred meters.

Commandant Henri Bonel, who headed the commission sent to Dayton in 1906 to negotiate with the Wrights, had been disappointed when the negotiations collapsed. Learning that the government was once more planning a purchase, Bonel was elated. By chance he ran into Fordyce and told him the good news. Fordyce went immediately to Letellier, who went to General Picquart and showed him a letter from Picquart's predecessor to the effect that all negotiations with the Wright brothers were to be made through Letellier. When Picquart told Deutsch that Letellier had prior right, Deutsch became angry with Wilbur.

Fordyce came to Berg with a proposal from Charles Humbert, member of the French senate and secretary of the budget committee of the French government, that Wilbur should ask 1,250,000 francs for his machine instead of the previously proposed 1,000,000. But when Berg learned that the extra 250,000 francs was a commission for negotiating the sale, he said he knew that Wilbur would not approve nor would he work with anyone in the government who expected to be paid in that fashion. Berg also told Humbert that if Humbert would not present the proposal to the Minister of War he would send it in another way and if no action was taken at once, he and Wilbur would leave for Berlin.

Wilbur was having trouble not only with the French government and men who wanted to help sell a machine. "I have not been able to get back to London where I ordered a dress suit," he wrote to his father. "I have ordered another one here and also a Prince Albert. They will cost about a hundred dollars. Please have Orville send me an American Express money order for $150. I cannot hobnob with the Emperor when I go to Berlin without some clothes. We are expecting to go there within a few days."

Orville to Wilbur (cablegram)

July 1, 1907

Not approve offer to French war dept. Have not received any information from Flint & Co. Do nothing without I consent. Keep me informed.

Wilbur received Orville's cable the morning of July 2. At noon the French government called off all negotiations.

Wilbur to Orville (cablegram)

Paris, July 1, 1907

Every trade is off. Send specifications of what is limit German government. When can you come?

Wilbur to Orville (cablegram)

July 10, 1907

Do you veto offers to French war department? When will you be able to ship machinery? Cannot get an answer.

Orville to Wilbur (cablegram)

Dayton, July 10, 1907

Left to discretion of you. Will be able to ship in one week. Can we do the work in France winter? Do not begin new negotiations till we have an understanding with Flints.

Orville wrote a stiff letter to Wilbur complaining that he was very much opposed to doing business with France, that he was disgusted with the Flints, that he did not approve of Wilbur's offer to the French and that he did not know what was going on. "When you cable you never explain anything so that I can answer with any certainty that we are talking about the same thing.

"You have not answered a single question I have asked in my cables. I did not answer your first cable the other day asking whether you should withdraw the French proposition, for the reason that I did not understand what you were talking about, for you asked in the next sentence when could I ship a machine. I could not understand why you were asking these things in the same cable as neither had any bearing on any previous correspondence."

Wilbur to Orville

Paris, July 13, 1907

Get help. Ship by express to N.Y. Cable when it has gone forward. You and Charlie come here as soon as possible. Expect to close.

Orville left for Paris July 20. On the same day Wilbur received a letter from his father complaining that his letters were not satisfactory and his cablegrams not explicit. Wilbur replied:

You people in Dayton seem to me to be very lacking in perspicacity. When the telegram came from Flint asking one of us to go to Europe at once, I saw instantly what was involved, and asked Orville to go. I did this for two reasons: (1) Because I wish the job of putting the final touches to the engines, and preparing the machine for shipment.

I am more careful than he is, at least I think so. (2) Because it was evident that the man who went to Europe would have to act largely on his own judgment without much consultation by letter or cable. I felt I was more willing to accept the consequences of any error on his part than to have him blaming me if I went. I never for a minute was so foolish as to suppose that the final decision should be made by the man at home, who, from the nature of the case, would necessarily be less competent to form a sound judgment than the man at the seat of the action. When Orville insisted that I should come, I told him I must be free to settle matters to the best of my judgment as they arose, as it would be impossible to wait for letters to pass back and forth or to explain things clearly by cable. Opportunities must be seized as they pass, or they are gone beyond recovery.

In my cables I always included every essential point on which his approval was desired. I made the questions as general as possible so that no understanding of details would be necessary. I myself saw to it that I kept the details substantially within the limits imposed by my question. No other course was possible. It has been no pleasure to me to assume to decide all these questions myself but it could not be helped. If positions had been reversed, I would have given him a free hand till I could have followed him to Europe myself; but I would have been ready to follow within one month instead of two months.

His complaint about not being able to find things is due to no fault of mine.

I wanted to attend to that end of the job myself, but he happened to be in one of his peculiar spells just then, and I soon saw that he was set on finishing the machine himself. So far as his letters indicate he spent his time on things of no use in the present situation, and left the necessary things undone.

Well, I will stop complaining. I suppose that he has been so worried that he has not really been himself. But it would have been much better if he had attended energetically to his department, and avoided interference in mine. I am sure that I have made no serious mistake and that I have done nothing that he would not have agreed to if he had been here. If a serious mistake has been made it lies in the assumption that the machine would be available quicker than now seems possible. I am not to blame for this.

Orville arrived in Paris July 28. When the porter who carried his trunk to a cab held out his hand, Orville tried to explain he had no French money. He handed the porter a dime. The man released a string of French. Assuming that perhaps he had not given enough, Orville handed him another dime. With another cascade of French, the porter gave him back his money. Orville never did understand what was wrong, he said in a letter to Katharine.

Leaving Orville to negotiate in Paris, Berg and Wilbur went to Berlin because the Germans had asked for a proposal. Wilbur suggested 100,000 marks ($25,000) for the first machine, 50,000 marks for training, a royalty of 30,000 marks each on the first ten machines and a lesser amount for further sales as long as the German patents lasted. Then he telegraphed Orville, who answered that as long as the French were considering a proposal which contained an exclusive clause, Wilbur should hold off. Wilbur had been in Berlin about ten days when Orville sent him a copy of the latest French proposal. Feeling that it was unsatisfactory, Wilbur returned to Paris to change the basis of negotiations or to withdraw. To his new proposals, which were similar to the ones he had made to the Germans, Fordyce made so many objections that the brothers withdrew all offers and Wilbur returned to Berlin to continue negotiations there.

Flint wrote that something was brewing in London. Orville went to London and found that the Barnum and Bailey Circus wished to hire the Wrights to give exhibition performances, but Wilbur wrote that they should do nothing about the matter until their other plans were settled.

The brothers spent part of August in Paris, where Orville injudiciously wrote to his father, "We have been real good over here. We have been at a lot of churches and haven't got drunk yet!"

Wilbur felt it necessary to write to his father four days later, "As to drinking and dissipation of various kinds you may be entirely easy. All the wine I have tasted since leaving home would not fill a single wine glass. I am sure that Orville and myself will be careful to do nothing which would disgrace the training we received from you and from mother."

The same day he wrote to Katharine, "I am sorry that you could not have been here all summer prancing up and down the Champs Elysees and rue de Rivoli and through the Louvre with red Baedeckers in hand, like the other teachers I have seen. It is so rich to see that look of ravenous thirst for knowledge; and combined expression of heavenly satisfaction and sore feet. Ha! Ha!"

Bishop Wright wrote to Wilbur September 5, "I did not anticipate that either of you would become intemperate or debauched, but I want you to show the foreigners that you are teetotalers, and in every way maintain that high character which it is most proper to have, and which in the eyes of the best in America is the most approved."

Although negotiations were under way in Russia, England, France and Germany, no definite proposals from any government had been made by late October. Wilbur wrote to Katharine October 22 that he doubted whether an agreement would be reached before they had made some demonstration flights and stirred up some excitement. "One of the German papers had a cartoon on us a few weeks ago, that is about the best thing we have seen. The Gebrüder Wright are represented as bargaining over the sale of a 'cat in a bag.' Orville is at one end bargaining with France while I am working Germany at the other end. France has a wheelbarrow full of money and is down on its knees begging us to accept it. The pile is marked 3,900,000 francs. Orville with a pipe in his mouth leans indifferently against the bag containing the wonderful machine, and with a bare glance at the 3,900,000 francs holds up four fingers to indicate 4,000,000 frs. is our bottom price. At the other end I am almost equally indifferent, though Russia is represented as pulling its last ruble out of its pocket.

"As soon as France scrapes up another 100,000 frs. we will sell out and come home."

Orville received in London late in October a letter from the Board of Ordnance, War Department, indicating the Board was interested in a proposal from the brothers. Orville replied that the business would require the brothers to be in Europe in the spring of 1908, but that if arrangements could be made for doing business with the government before spring, one of them would return to the United States and arrange to make demonstrations in one of the southern states. "Nothing would give us greater pleasure," he said, "than to furnish the *first* machine to it (our own government)."

They signed an agreement with Flint and Hart Berg to act as their sole agents abroad in negotiating agreements with governments and in forming companies to take over promotion of the Wright invention. Wilbur left for the United States November 11. Orville remained behind to talk with a company that was going to make a bid to build engines for them during the winter.

Wilbur stopped in Washington and talked with General William Crozier, General James Allen and Major Lawson Fuller. They told him that $10,000 was available for a machine. Wilbur said that the price was $25,000 and went home.

He returned to Washington December 5 for a meeting of the Board of Ordnance. He offered a machine capable of carrying two people for $25,000. Again the Board made no offer, but General Allen, Chief Signal Officer, United States Army, said that shortly the Board would advertise for bids on flying machines. "When I first learned that the Board was advertising for bids," Wilbur wrote to Chanute on his return to Dayton, "I doubted its good faith, but am now inclined to think I did them an injustice in suspecting such a thing. On the whole, I think the conditions are fair, though a limitation of three trials is rather severe."

Orville returned home in December and the family celebrated Christmas dinner at Lorin's home, where the talk was mostly of the wonderful sights the brothers had seen in Europe. Although the year 1907 ended without the sale of one flying machine, the brothers were not discouraged. Wilbur reminded Chanute that he had earlier predicted that it would take at least five years for an independent solution to the flying problem and said that he still believed it. "The two years that have passed since Archdeacon, Santos, and Ferber predicted that the feats of the Wrights would be surpassed within three months have seen all other predictions than ours overturned or repeatedly amended. I have confidence that our prediction will stand solid after the scythe of time has reaped several crops of French predictions."

The brothers looked forward with joy to the year 1908. Fortune and fame, they felt, would be theirs.

On the launch rail at
Ft. Myer, Virginia, 1908

9

Infinite Highway of the Air

*But it is not really necessary to look too far into the future;
we see enough already to be certain that it will be magnificent.
Only let us hurry and open the roads.*

—Wilbur Wright, 1908

When the Signal Corps advertised for bids for military flying machines, newspapers and magazines country-wide editorially hooted at the idea. "There is not now a known flying-machine in the world which could fulfill these specifications at the present moment," declared the *American Magazine of Aeronautics.* "Had an inventor such a machine as is required would he not be in a position to ask almost any reasonable sum from the government for its use? Would not the government, instead of the inventor, be a bidder?"

The Signal Corps received forty-one bids. Only three were accompanied by the check stipulated as a condition of bidding. One bid was from J.F. Scott for $1,000 with delivery in 185 days. Another bid came from Augustus Herring for $20,000 with delivery in 180 days, and one came from the Wrights for $25,000 with delivery in 200 days.

The government accepted all three bids. Scott immediately withdrew. Herring tried to sublet his contract to the Wrights, but he had the same success as when he proposed to give them two-thirds ownership of their own Flyer.

Alexander Graham Bell, Glenn Curtiss, Thomas Baldwin, and Lieutenant Thomas Selfridge, U.S.A., among others had organized the Aerial Experiment Association with headquarters first in Nova Scotia and later at Hammondsport, New York, home of Curtiss. Selfridge, secretary of the association, wrote to the Wrights January 15, 1908, asking advice on glider construction, particularly how to make the ribs strong enough to maintain their curvature under ordinary circumstances and also asked about a good way to fasten them to the cloth and upper lateral cords of the frame.

Wilbur, never doubting that Selfridge wanted the information for anything except scientific purposes, answered fully and suggested that Selfridge read his two addresses to the Western Society of Engineers, Chanute's article in *Revue des Sciences,* and their own patent number 281,393.

Chanute wrote to Wilbur January 19, commenting that he thought it unfortunate that the Wrights had not organized a sustaining company at the beginning. He believed that the eventual financial result to them depended upon "the unpleasant eventuality of serious accidents to some of your competitors."

Wilbur disagreed. He said that the failures of the French flyers made trouble for the Wrights by throwing more and more doubt on the practicability of any aeroplane. He also disagreed with Chanute that he and Orville had made a mistake in failing to organize a company at the beginning. "It is true that we could have made big money by selling stock to ignorant gullible investors, but we did not care to do that. As for selling to real businessmen on a strictly business basis, we feel certain that we can do much better today than we could have done at the beginning. In organizing a company the patents are the important thing. They are as fully valuable now as at the beginning."

Flint and Company asked Wilbur to go to New York March 15 because a group of French capitalists including Lazare Weiller and Deutsch had agreed to buy the Wrights' French patents after a series of demonstrations. Payment would be 500,000 francs with the delivery of the first machine, 50 percent of the founders' shares of stock, and 20,000 francs for four more aeroplanes. Wilbur accepted the offer.

Now that the Wrights had contracts with the United States War Department and with the French company, they wanted to modify and improve their aeroplane before making demonstration flights. They changed the 1905 machine so that the operator and one passenger could sit upright rather than lie in a prone position. Both operator and passenger had controls. The aeroplane, therefore, could be used to train pilots.

Wilbur took the altered 1905 machine to Kitty Hawk to practice for the Signal Corps tests. He wrote to Chanute that he wished the European sales could have been put off until the United States business was over, because he and Orville worried that one of them would have to go to France before the Signal Corps tests began.

When Wilbur arrived at Kitty Hawk April 10 he found things "pretty well wrecked," he said in his diary. "The side walls of the old building still stand but the roof and north end are gone. The new building is down and torn to pieces. The pump is gone."

Bad weather and slow delivery of supplies prevented renovation of the camp for weeks. By the time Orville arrived on April 25 Wilbur had a regular camp bed, but Orville had to sleep on a board thrown across the ceiling joists, while Charlie Furnas, a Dayton mechanic, had to sleep on the floor.

They began practice with the altered machine May 6, 1908. May 12 they received a letter from Katharine enclosing a telegram from Berg in France saying that Weiller refused to postpone the French tests. Wilbur wired, asking for a ten-day extension. (The agreement with Weiller was that the first demonstration flight must be made within four days of the time designated for the first attempt and no later than five months from the first of June.)

Practice flights continued. Whereas in the past the brothers had been left strictly alone, now the newspapers sent reporters and photographers to follow every move. Sometimes the brothers could see them hiding in the woods or behind sand dunes.

Bishop Wright's Diary

May 9, 1908

Last evening's *Herald,* today's *Journal,* the *Cin. Commercial,* the *Cin. Enquirer,* and *Dayton News,* had alleged dispatches concerning the Wright brothers' flights, the *Enquirer* reporting they had flown 3,000 feet high, 30 miles long, and 8 miles out to sea!

Wilbur and Charlie Furnas made the world's first flight carrying two men May 14, 1908. The flight of 600 meters was made in 28 and three-fifths seconds.

Wilbur in his usual business suit and cap checks the French-built Wright A machine at Hunaudières Race Course, five miles south of Le Mans.

A little later in the morning Orville and Charlie made a short flight. On the second lap around the West Hill the engine overheated and they came down. After dinner Wilbur went for a flight alone. "I passed over dry ponds and, passing camp, proceeded around the West Hill and the Little Hill by the sound and was continuing on the course followed on the first round when the machine suddenly darted into the ground when going with the wind at a velocity of about 85 kilometers an hour. The front frame and upper surface were wrecked. The front rudder & tail & machinery and the wing sections were almost intact. The lower center section was only slightly injured. I was thrown violently forward and landed against the top surface but remained inside the body of the machine. I received a slight cut across the bridge of my nose, several bruises on my left hand, right forearm, and both shoulders. The next day I felt a little stiff all over. We tore the machine apart and put the lower surface on our truck with the engine in place, and piled on the transmission & a few other parts, between four and five hundred pounds in all, and the three of us dragged it back to camp, a distance of about a mile and a quarter. The heat had become almost unbearable and we barely escaped collapse before reaching the camp. After dark we went over and got the rudders and radiators. We went to bed completely fagged out."

The next day despite the intense heat they went to the scene of the wreck and hauled back the wing tips and uprights. Two photographers, one from the *Chicago Tribune,* one from the *London Daily Mail,* arrived. "They wished," Wilbur noted in his diary May 15, "to take a picture of Orville in his 'Merry Widow' bonnet and me in my dog harness hitched to the trucks, but finally desisted. It would have been an amusing picture for private use, but not such as we cared to have spread broadcast."

"In the afternoon we received telegram from Flints saying that ten-day postponement had been granted by Weiller and French business. Soon after noon the wind suddenly shifted to the north and soon became uncomfortably cold."

Wilbur left camp for France May 18 and arriving in New York May 19, he found that Katharine had sent a trunk and hatbox to his hotel. "I do sometimes wish though that you had raised the lid of my hatbox, which was not locked," Wilbur wrote to Katharine, "and put some of my hats in it before sending it on. However, a man can buy hats almost anywhere."

Wilbur to Orville

New York, May 20, 1908

I stopped a few moments at the *Century* office and told them I was going abroad, and that it was probable that we would be doing business publicly before an article could be furnished them as heretofore talked of, but that if they wished they could communicate with you in regard to getting something ready, but intimated that you probably would not have time to do so now. However, it is my opinion as firmly as ever that we need to have our true story told in an authentic way at once and to let it be known that we consider ourselves fully protected by patents. One of the clippings which I enclose intimates that Selfridge is infringing our patent on wing twisting. It is important to get the main features originated by us identified in the public mind with our machines before they are described in connection with some other machine. A statement of our original features ought to be published and not left covered up in the patent office. I strongly advise that you get a stenographer and dictate an article and have Kate assist in getting it in shape if you are too busy.

He also suggested that Orville should come to France in a few weeks and practice flying there before one of them had to go to Ft. Myer for government trials. "This plan would put things to the touch quickly, and also help ward off an approaching financial stringency which has worried me very much for several months."

Wilbur sailed for France May 21, arriving May 29. Léon Bollée, a factory owner, invited Wilbur and Hart O. Berg to Le Mans to inspect grounds for practicing, training flyers, and giving demonstrations. Wilbur liked the Hunaudières Race Course and the fact that Bollée offered him the use of his factory. Berg rented the race course. Wilbur selected a place for a building to house the aeroplane.

Léon Bollée drives Wilbur and three others and hauls the Model A Flyer from Hunaudières to Camp d'Auvours, several miles east of Le Mans, August 18, 1908.

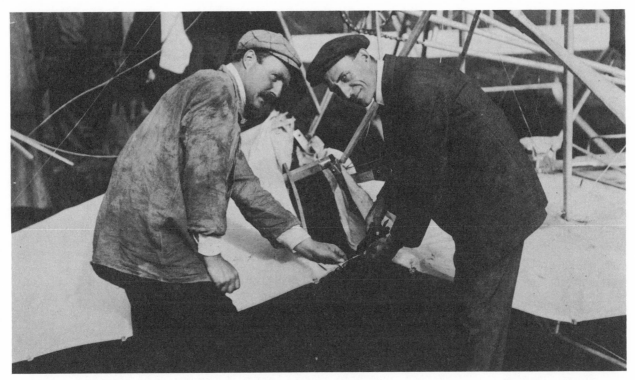

*Hart Berg's chauffeur
Fleury and Wilbur strike
a pose following an
accident at Hunaudières
August 13, 1908.*

Wilbur to Orville

Le Mans, June 17, 1908

I opened the boxes yesterday and have been puzzled ever since how you could have
wasted two whole days packing them. I am sure that with a scoop shovel I could have
put things in within two or three minutes and made fully as good a job of it. I never saw
such evidences of idiocy in my life. Did you tell Charley not to separate anything lest it
should get lonesome? Ten or a dozen ribs were broken and as they are scattered here and
there through the surfaces it takes almost as much time to tear down and rebuild as if we
could have begun at the beginning. One surface was so bad that I took it completely
down. Never again pack anything else in the surface box. The cloth is torn in almost
numberless places and the aluminum has rubbed off of the skid sticks and dirtied the
cloth very badly. The radiators are badly mashed; the seat is broken; the magneto has
the oil cap broken off, the coils badly torn up, and I suspect the axle is bent a little; the
tubes of the screw support are mashed and bent. The only thing I ever saw resembling
the interiors of the boxes in the rattler at a foundry. Please bear in mind hereafter that
everything must be packed in such a way that the box can be dropped from a height of
five feet ten times, once on each side and the other times on the corners. The boxes must
be cleated outside and the lids must be put on with screws as they must be opened by the
customs authorities. Such things as magnetos and other articles of similar size must not
be thrown loosely into a box fifteen feet long, but must be put into medium-size boxes
and enclosed in the bigger boxes if desired. To be brief, things must be packed at least
ten times as well as they were last time. And everything must be listed and the net
weights taken of the stuff in each box.

While Wilbur was working on an engine at Le Mans July 4, a rubber tube came off
the upper water connection with the engine. The boiling water struck him on the forearm, chest, and side. "M. Bollée had some picric acid on hand and at once applied it.
My escape from more serious consequences was doubtless due to this prompt

1907 Machine
The 1907 machine, flown in 1908 at Ft. Myer and in France and in Germany and in Italy in 1909, had a wingspan of 41 feet, a wing area of 510 square feet, a horizontal rudder area of 70 square feet, a vertical rudder area of 23 square feet. It was about 31 feet long and weighed about 800 pounds.

treatment," he wrote his father. "The blister on my arm was about a foot long and extended about two-thirds of the way round my arm. That on my side was about as large as my hand."

Wilbur to Katharine

Le Mans, July 7, 1908

In order to avoid any chance of my arm getting sore, I had a "docteur," probably a "hoss" doctor, come to dress my arm Sunday. He sent for a bale of cotton and a keg of oil, and after soaking the former in the latter made a vain attempt to plaster it on to my arm and side before more than half the oil had dripped out. When he was done he had two wash bowls, six towels and a dozen or two newspapers soaked with oil, not to speak of the table cover, the rug and my clothes. The oil ran down my arm and began dripping off my finger tips and down my legs till my shoes were half full. As I had no tin handy to make eave troughs I got a dozen newspapers and spreading them on the bed tried lying down. But the oil went through all the newspapers, the sheet and into the mattress. I thereupon removed all the stuffing, like the fat man in A. Ward's show, and dressed the burns myself with more sense.

I fired the "docteur" after his first visit. If you ever get burned do not waste your money on doctors, but get a barrel of oil and fill up your bath tub and crawl in and stay till you are well.

Back in Dayton Orville read in the *Scientific American* that Glenn Curtiss was using movable surfaces at the tips of the wings of his *June Bug*. The tips were adjustable to different angles for maintaining lateral balance. Orville wrote to Curtiss July 20, "In our letter to Lieutenant Selfridge of January 18th replying to his of the 15th in which he asked for information on the construction of flyers, we referred him to several publications containing descriptions of the structural features of our machines, and to our U.S. Patent No. 821,393. We did not intend, of course, to give permission to use the patented features of our machine for exhibitions or in a commercial way." Orville pointed out that the patent specifically covered the combination which they were informed Curtiss was using. "If it is your desire to enter the exhibition business," Orville said, "we would be glad to take up the matter of a license to operate under our patents for that purpose."

Léon Bollée's photo album shows this record of a visit by Orville and Wilbur [at extreme right] to the Bollée home. Bollée is at left, and Hart Berg stands to the right of Bollée's daughter and wife.

Wilbur grasps controls for first flight in Europe August 8, 1908. Wilbur's right hand when moved fore and aft controls the vertical rudder and when moved side to side controls wing warping for lateral balance. Wilbur's left hand moves the left stick forward and back, which moves the front elevator [which controls the pitch].

Curtiss replied that he did not intend to enter the exhibition business and that he had referred the matter of patents to Lieutenant Selfridge, secretary of the Aerial Experiment Association.

Wilbur made his first European flight at Hunaudières Race Course August 8. It was witnessed by a large crowd, which included members of the Aéro-Club de France and members of the press. He made the flight of one minute, forty-five seconds using stick control for the first time. On the second flight he made a figure eight and landed at the starting point. "The newspapers and the French aviators nearly went wild with excitement," he told Orville in his next letter. "Blériot & Delagrange were so excited they could scarely speak, and Kapferer could only gasp and could not talk at all. You would have almost died of laughter if you could have seen them. You never saw anything like the complete reversal of position that took place after two or three little flights of less than two minutes each. Deutsch telegraphed to inquire whether he could have the 100,000 frs. stock and definitely took it. The English Mercedes-Daimler Co. have written to know whether they can have England on same terms as the published Weiller contract. They also would like to arrange the German business, I presume through the German Daimler Co. I have asked them to send a man to talk over matters.

It is a cold day at Camp d'Auvours, 1908, and Wilbur wears two coats.

The French cartoon caption reads, "L'Aviation, a ses petits ennuis. Le Mans— Camp d'Auvours, 1908."

A derrick, a heavy weight, and a system of pulleys catapult the machine into the air. Wilbur circles the field at Camp d'Auvours, 1908.

"How fast is the wind blowing?" Orville asks Wilbur at Pau. Wilbur checks with the Richard anemometer.

Just the thing to wear at a picnic. Orville in bowler and overcoat, Katharine in hat and walking suit, and oxen in working garb, Pau, 1909

The future has arrived; only the oxen do not know. Pau, 1909

"We certainly cannot kick on the treatment the newspapers have given us; even *Les Sports* has acknowledged itself mistaken."

After nine flights at Hunaudières Wilbur moved to a larger field at Camp d'Auvours and put up a new shed. There he made a series of flights through December 31, 1908. He carried his first European passenger, Ernest Zens, a French balloonist, September 16, the same day he set a new French distance record of 39 minutes, 18 and 2/5 seconds. "An artist has been around the place trying to make a portrait of me for one of the big illustrated papers," he wrote Katharine. "I have given him a few standings but I have been too busy for sittings. He will go to America to get one of Orville to accompany it. He has tried to catch that enigmatic smile which the papers talk so much about, but I fear you will raise an awful fuss when you see it. It is worse than shirt sleeves! We will soon have a picture you certainly cannot kick about on the score of shirt sleeves. Yesterday it was rather cold and to shut out the winds during the flight I slipped on a blue jumper or overhaul jacket. It left about two inches of my coat sticking out of the bottom. After the flight Mr. and Mrs. Weiller & Mr. Deutsch came up to congratulate me and the *Vie au grand air* photographer slipped up and snapped us. Mrs. Weiller laughed heartily when I told her how you objected to shirt-sleeve pictures and how pleased you would be to see that I had on two coats."

In the same letter to Katharine Wilbur said that Berg had started to Vienna "but when he got as far as Paris and heard that I had made a record flight of about 20 minutes, he came chasing back, although he had promised his wife, who was near Vienna, that he would bring her home. Mrs. Weiller reports that Mrs. Berg is raking Berg over the coals in great shape, and demanding whether he prefers Wright to herself. Mrs. Berg has many redeeming qualities and is really a charming woman, like yourself. (I put my head under the table while I wrote this—instinct I suppose.) So goodbye for the present."

Orville left for Washington on his thirty-seventh birthday, August 19, 1908, after shipping the Wright 1908 machine to Ft. Myer for the Signal Corps trials. After working two weeks assembling the machine, he made his first flight September 3, the first of 14 flights extending through September 17. On September 9 he established three new records. He made a flight of 57 minutes, 31 seconds, circling the field 57 times, a second flight of 62 minutes, 15 seconds circling the field 55 times, and a flight of 6 minutes, 24 seconds with Lt. Frank P. Lahm as passenger. On September 10 he flew 65 minutes, 52 seconds at an estimated altitude of 200 feet. The next day, September 11, Orville flew 70 minutes, 24 seconds, circling the field 57½ times, making two figure eights, the first seen at Ft. Myer. September 12 he set two new records: with Major George O. Squier as passenger he flew 9 minutes, 6 and 1/3 seconds. He then made a new distance record, circling the field 71 times, flying 1 hour, 14 minutes, 20 seconds.

Orville had written to Wilbur a few days before that Selfridge was at Ft. Myer and that he didn't trust him an inch. "He is intensely interested in the subject, and plans to meet me often at dinner, etc. where he can try to pump me. He has a good education, and a clear mind. I understand that he does a good deal of knocking behind my back."

Lieutenant Selfridge asked to fly as a passenger with Orville September 17. They took off without incident and made three rounds of the field. On the fourth round Orville made a wider circuit with less abrupt turns. Trouble began on the first slow turn. While flying towards Arlington Cemetery, Orville heard a light tapping in the rear of the machine. He looked quickly around but saw nothing wrong. He decided, however, to shut off the power and descend as soon as the machine could be turned in a direction for

September 3, 1908, Ft. Myer, Virginia: the 1908 Flyer is ready for its trial. The photo shows the bicycle-chain transmission, engine, fuel tank, two seats, and control sticks.

landing. He heard two big thumps and the machine began to shake. The machine suddenly veered to the right and Orville immediately shut off the engine.

The machine would not respond to the steering and balancing levers. While he continued to push the levers, the machine turned to the left till it faced the field directly. He tried to stop the turning by reversing the levers and to level the wings, but the machine turned down and headed straight for the ground. Neither man spoke, but when the propeller broke, Selfridge looked inquiringly at Orville. As the fall began, Selfridge exclaimed, "Oh! Oh!"

Orville kept working the levers. Within about twenty-five feet of the ground, the machine began to right itself rapidly. "A few feet more," Orville said later, "and we would have landed safely. As it was, the skids hit out at the front end. All the front framing was broken."

Lieutenant Selfridge was fatally injured, and Orville sustained three hipbone fractures, a dislocation of one hip, a fractured left leg, and four broken ribs.

When Katharine heard the news of the crash, she left at once for Washington. Charles Flint arrived about the same time and the two went together to the hospital. Katharine announced she was staying until Orville was able to return home.

September 17, 1908, Ft. Myer, Virginia: Lt. Thomas E. Selfridge and Orville prepare for take-off.

Wilbur to Katharine

Le Mans, September 20, 1908

I received the news of the awful accident at Washington just as I was finishing preparations for an official trial for the Michelin & Commission d'Aviation prizes. The death of poor Selfridge was a greater shock to me than Orville's injuries, severe as the latter were. I felt sure "Bubbo" would pull through all right, but the other was irremediable. The weather was ideal, a day of a thousand, but in view of the positive news of Selfridge's death, I did not feel that it would be decent to proceed as though I were indifferent to the fate which had befallen him as a result of his trust in our machines. So the trials were postponed till next week.

I cannot help thinking over and over again "If I had been there, it would not have happened." The worry over leaving Orville alone to undertake those trials was one of the chief things in almost breaking me down a few weeks ago and as soon as I heard reassuring news from America I was well again. It was not right to leave Orville to undertake such a task alone. I do not mean that Orville was incompetent to do the work itself, but I realized that he would be surrounded by thousands of people who with the most friendly intentions in the world would consume his time, exhaust his strength, and keep him from having proper rest. When a man is in this condition he tends to trust more to the carefulness of others instead of doing everything and examining everything himself.

I cannot help suspecting that Orville told the Charleys to put on the big screws instead of doing it himself, and that if he had done it himself he would have noticed the thing that made the trouble, whatever it may have been. If I had been there, I could have held off the visitors while he worked or let him hold them off while I worked. But he had no one to perform this service. Here Berg helps to act as a buffer and gives me some chance to be alone when I work. People think I am foolish because I do not like the men to do the least important work on the machine. They say I crawl under the machine and over the machine when the men could do the thing well enough. I do it partly because it gives me opportunity to glance around to see if anything in the neighborhood is out of order. Hired men pay no attention to anything but the particular thing they are told to do, and are blind to everything else.

Tell "Bubbo" that his flights have revolutionized the world's belief regarding the practicability of flight. Even such conservative papers as the London *Times* devote leading editorials to his work and accept human flight as a thing to be regarded as a normal feature in the world's future life.

Orville has a way of stepping right into the affections of nice people whom he meets, and they will be nice to you as first for him and then for yourself, for you have some little knack in that line yourself. I am glad you are there to keep your eagle eye on pretty young ladies. I would fear the worst, if he were left unguarded.

I presume that poor old Daddy is terribly worried over our troubles, but he may be sure that things will turn out all right at last. I shall be not only careful and more careful, but also more careful, and cautious as well. So you need have no fears for me.

Just after the accident when the Charleys were packing the wrecked machine to ship it back to Dayton, Katharine heard that Alexander Graham Bell had examined the machine. Knowing that Dr. Bell was the founder of the Aerial Experiment Association, Katharine asked Chanute, who had come to Washington to visit Orville, to find out what he could. After making an investigation, Chanute said that almost everything was already packed when Bell arrived and all that Bell had done was whip out a tape and measure the wing width. Neither Katharine nor Orville believed that Bell's measuring the wing was an innocent act.

September 17, 1908: Lt. Selfridge dies in crash at Ft. Myer. Orville suffers severe injuries. The crash ends the Wright Army trials for 1908.

Orville remained in the hospital in Washington until the first of November. He went home to Dayton and wrote to Wilbur, "The only explanation I have been able to work out of the cause of the plunge for the ground is that the rear rudder, after the stay wire was torn loose by the propeller, fell over on its side and in some mysterious manner was caught and held in this position, with a pressure on its under side."

"After looking over the Ft. Myer machine," Wilbur wrote to Chanute after he returned to Dayton, "we have decided that the trouble came in the following manner. One blade of the right propeller developed a longitudinal crack which permitted that blade to flatten out and lose its pushing power. The opposite blade not being balanced by an equal pressure on the injured blade put strains on its axle and its supports which permitted it to swing forward and sidewise a little farther than the normal position and at the same time set up a strong vibration. This brought the uninjured blade in contact with the upper stay wire to the tail and tore it loose, the end of the wire wrapping around the end of the blade and breaking it off.

"Now that we have located the trouble we are certain that its recurrence can be avoided. We are making a set of screws which are heavier at the weak point and will strengthen them with canvas all the way down the concave side. Then we will brace the axle supports in such a way that no amount of vibration can enable the screws to reach the tail braces. I am glad it was no carelessness of Orville that brought about the catastrophe. It is so easy to overlook some trifling detail when setting up a machine under the conditions which existed at Ft. Myer, that I feared he might have failed to properly secure a nut somewhere; but everything was found tight after the accident except the fastening which was torn out when the screw struck the tail stay wire. In so far as the responsibility falls upon anyone I suppose it falls upon me as I did the testing of the screws. I see now that a better test would have been to turn the full power of the motor on to a single screw and thus give it an over-strain under conditions similar to those it undergoes in use. It is the first time we have ever had any indication of trouble with our propellers."

While Orville was recovering in the Washington hospital, Wilbur resumed his flying in France. On September 21 he established a new world flying record of one hour, 35 minutes, 24 4/5 seconds. A week later Wilbur competed for a 5,000-franc prize offered by the Aéro-Club de France, flying one hour, seven minutes, 24½ seconds with a maximum altitude of ten meters. Wilbur won.

The Aéro-Club de la Sarthe honored Wilbur at a reception for establishing this record. Asked to say a few words, Wilbur responded, "I know of only one bird, the parrot, that talks, and he can't fly very high."

Mrs. Hart O. Berg was the first woman to fly as a passenger when Wilbur took her for a flight October 7, 1908. Mr. Berg tied a rope around his wife's skirt near her ankles to keep her skirt from flying in the wind. When she landed and walked away from the machine with the rope still in place, she was able to take only mincing steps. In the crowd was a designer from the House of Paquin. He made notes and rushed to fashion a dress with a skirt so narrow at the ankles that the wearer had to make the same mincing steps. It became known as the hobble skirt.

A French actress, Mme. Sorel, needed a costume for a role in which she had to stand against a pillar for a long time. The director wanted her figure to contrast with the pillar and yet be harmonious with it. Since the hobble skirt was ideal, she wore one. Thereupon the fashionable women of Paris bought hobble skirts and soon all over the world women were hobbling. The style lasted for several years, not only for leaning against pillars, but for walking and even dancing.

September 9, 1909: Mrs. Alfred Hildebrandt in hobble skirt talks with Orville. She is the first woman to fly as a passenger in an aircraft in Germany.

The day after Mrs. Berg's flight, Wilbur took his first English passenger for a flight. He was Griffith Brewer, one of the first members of the Aeronautical Society of Great Britain, to see Wilbur fly. In fact, Brewer hoped that he would be the first, but as he was leaving for France he happened to see Charles Rolls, founder of the Rolls-Royce motor car firm. When Brewer confided to Rolls that he was on his way to see the flight, Rolls laughed and said he had just come from there. The Brewer flight was made in the presence of the Dowager Queen Margherita of Italy.

Flying with Paul Painlevé of the French Academy of Sciences as a passenger on October 10, Wilbur made a world record for duration—one hour, nine minutes, 45 and 2/5 seconds—a flight which officially fulfilled the performance conditions of their contract with Lazare Weiller.

Wilbur described the flight in a letter to Orville and mentioned a number of prizes he intended to compete for, one for a flight across the English Channel and another prestigious one, the Michelin Prize for 1908. "Maxim was here several days this week," he went on. "I doubt the goodness of his purpose and dislike his personality. He is an awful blow and abuses his brother and son scandously. Herr Loewe was also here. He offered $25,000 and ¼ interest in the business for our German interests. I merely smiled. Did I tell you I had a letter of congratulation from the German Crown Prince last week? I suspected at first that it was a fake, but it seems to have been genuine.

"I will send some money home soon. Have Kate get a year's leave of absence and take a rest. You ought to write a letter to Léon Bollée as president of the Aéro-Club of the Sarthe, thanking it for its telegram to you on the evening they met to drink our healths, on the occasion of my putting up the record of an hour and a half. They will appreciate the courtesy. Do not neglect it."

Bishop Wright's Diary

November 1, 1908

Orville and Katharine came home from Ft. Myer, Va., arriving at 9:00 a.m. He is brought out from the depot on a wheeled chair. His mind is as good as ever and his body promises to be in due time. Carrie Grumbach gets the dinner. Lorin met Orville at the depot and he and Nettie dine with us. A few call. Flowers for Orville and Katharine came in.

In France Wilbur received honors, citations, and gold medals nearly every day. On one day he was honored at a luncheon of the French Society *Autor du Monde,* attended by editors, authors, scientists, and statesmen. After the luncheon he was taken to the French Senate, which adjourned in his honor and held a reception. In the evening the Aéro-Club de France honored both brothers at a banquet, presenting Wilbur with the club gold medal and the *Académie des Sports* gold medal inscribed "To the conquerors of the air, M.M. Wilbur and Orville Wright, the first to fly with an apparatus heavier-than-air driven by a motor." In response Wilbur said:

For myself and my brother I thank you for the honor you are doing us and for the cordial reception you have tendered us this evening.

If I had been born in your beautiful country and had grown up among you, I could not have expected a warmer welcome than has just been given me. When we did not know each other, we had no confidence in each other; when we are acquainted, it is otherwise; we believe each other, we are friends. I thank you for this. In the enthusiasm being shown around me, I see not merely an outburst intended to glorify a person, but a tribute to an idea that has always impassioned mankind. I sometimes think the desire to fly after the fashion of birds is an ideal handed down to us by our ancestors who, in their gruelling travels across trackless lands in prehistoric times, looked enviously on the birds soaring freely through space at full speed above all obstacles, on the infinite highway of the air. Scarcely ten years ago all hope of flying had almost been abandoned; even the most convinced had become doubtful, and I confess that, in 1901, I said to my brother Orville that men would not fly for fifty years. Two years later, we ourselves were making flights. This demonstration of my inability as a prophet gave me such a shock that I have ever since distrusted myself and have refrained from all prediction—as my friends of the press, especially, well know. But it is not really necessary to look too far into the future; we see enough already to be certain that it will be magnificent. Only let us hurry and open the roads.

Once again, I thank you with all my heart, and in thanking you I should like it understood that I am thanking all of France.

May 1, 1909: The Aero-Club de la Sarthe, Le Mans, Leon Bollée, president, presents a bronze sculpture by Louis Cardin, showing the Muse of Aviation embracing the "Freres Wright," points to the eagle, whose realm they have conquered.

10

Bittersweet

Every sweet has its sour; every evil its good.
—Emerson, *Compensation*

After Wilbur moved his flying activities to Pau, a resort in southern France, he went early in January 1909 to meet Orville and Katharine in Paris. On the day of their arrival the Automobile-Club de France gave a luncheon during which Andre Michelin presented Wilbur with the Michelin Award of 20,000 francs for his record-breaking flight of December 31. Michelin presented it in cash, a sizeable stack of bills. Wilbur took the stack in his hands, ran his thumb nail down the middle, divided the stack, put one half into his pocket and handed the other half to Orville.

From January to April at Pau life for the Wrights was one round of gaiety and adulation. Crowds rushed to the field every day to watch the flying. Lord Alfred North-cliffe, owner of the *London Daily Mail,* came as well as the former Prime Minister of England, Lord Arthur J. Balfour, Lord and Lady Frederick G. Wolverton, King Alfonso XIII of Spain, Edward VII of England, and assorted princes, princesses, dukes, duchesses, admirals, and generals. Katharine loved it. She had special dresses made for meeting royalty.

The Board of Regents of the Smithsonian Institution in February recommended that the newly established Langley Medal be presented to the brothers "for advancing the science of aerodynamics in its application to aviation by their successful investigations and demonstrations of the practicability of mechanical flight by man." The State Senate of Ohio and the Dayton City Council passed resolutions to present medals.

Congress by resolution in March awarded the Congressional Medal to the Wright brothers "in recognition of the great service of Orville and Wilbur Wright of Ohio rendered the science of aerial navigation in the invention of the Wright aeroplane, and for their ability, courage, and success in navigating the air."

While honors were being planned in the United States, the brothers moved their flying activities to Centocelle in Italy. There Wilbur had an audience with King Victor

Wilbur instructs student-pilot Count Charles de Lambert at Pau, 1909. Coachmen hold the skittish horses.

Orville Wright and Hart O. Berg talk with Crown Prince Friedrich Wilhelm, the first member of royalty to ride in an aeroplane. After the 1909 flight in Germany the Crown Prince gave Orville a jeweled stickpin, a crown set with rubies with the Prince's initial "W" in diamonds. Orville said later, "The 'W' stands for Wright."

"To make a left turn," Wilbur says, "move the right stick forward." His student is King Alfonso XIII of Spain. Pau, 1909. The King had promised the Queen and his Cabinet that he would not fly.

King Edward VII of England did not fly, either, but he watched two flights at Pau, 1909, one with Katharine as a passenger. Under the flowered hat behind the King is Katharine.

Emmanuel III of Italy. He carried in turn in his aeroplane several members of the royal family, military officers, ambassadors, and once a camerman who took the first motion pictures from an aeroplane in flight.

After a number of farewell dinners and receptions in Rome, Paris, and London, the Wrights left for the United States on May 5, 1909.

Bishop Wright's Diary

May 13, 1909

Flags, Chinese lanterns, and electric lights are being arranged. Eleven carriages met the Wright brothers and Katharine at the depot. And a four-horse carriage pulls them home, where thousands meet them around our house. Over 10,000 came that night. Fireworks ended the evening.

Dayton celebrates the brothers Wright for two days, June 17, 18, 1909, with fireworks, flags, flares, bunting, and electric lights. Monument Avenue and Main Street

Standing on either side of President William Howard Taft are Wilbur and Orville Wright holding their just-presented gold medals of the Aero Club of America, June 1910. Katharine wears a new hat and gown for the occasion.

The brothers went to Washington on June 10 where President William Howard Taft presented them with the Aero Club of America gold medals. Nearly one thousand persons attended the celebration in the East Room of the White House. President Taft said, ''You made this discovery by a course that we of America like to feel is distinctly American—by keeping your nose right at the job until you had accomplished what you had determined to do.''

Chanute wrote that he was delighted with all the good news of the Wrights. ''I have, of course, rejoiced over your triumphs in Europe and was particularly gratified with the sensible and modest way in which you accepted your honors, both abroad and since your return to this country. It encourages the hope that you will still speak to me when you become millionaires.'' He added that he was beginning to feel his advancing age and his capacity for work was diminished. He had bought a house for his daughters and hoped that the Wrights would visit him there.

Every bell in Dayton rang and every factory whistle blew for ten minutes at 9 a.m. June 17, 1909, to signal the beginning of a two-day celebration to honor the Wrights. A carriage preceded by bands took them to the opening events. Orville's boyhood friend, Ed Sines, and Edward Ellis, a friend of Wilbur, rode in the carriage with them to a review of an exhibition parade and drill by the Dayton Fire Department, a reception at the YMCA, and presentation of the key to the city.

In the afternoon of the first day the brothers went back to their shop to work, but callers all afternoon prevented much work. One of the visitors was James Whitcomb Riley. In the evening they attended a fireworks display climaxed with a set piece with their portraits in fireworks entwined with an American flag.

On the second day Bishop Wright gave the invocation at a great gathering at the fairgrounds. General James Allen, Chief of the Signal Corps, presented the Congressional Medal, Governor Judson Harmon presented the Ohio State Medal, and Mayor Edward Burkhardt presented the Dayton Medal. A living American flag of 2,500 Dayton schoolchildren dressed in red, white and blue sang patriotic songs.

Although the brothers had philosophically borne years of being ignored by nearly everyone, they found it more difficult now that fame had come their way. Celebrations such as Dayton had put on kept them from their work. Chanute said to them, ''You state that this last demonstration was in opposition to your own wishes. I know that the reception of such honors becomes oppressive to modest men and they would avoid them if they could, but in this case you have brought the trouble upon yourselves by your completing the solution of a world-old problem, accomplished with great ingenuity and patience at much risk of personal injury to yourselves, and I hope that when the present shouting is over you will continue to achieve further success and to receive ample rewards of all kinds.''

The shouting was barely over when they left for Washington to assemble their machines for the government tests. The flight July 27 with Lieutenant Frank P. Lahm fulfilled the Army requirement of flying for one hour, carrying two persons. President Taft, the Cabinet, and a crowd of ten thousand witnessed the two-man flight at Ft. Myer.

With Lieutenant Benjamin D. Foulois as passenger July 30 Orville flew over a measured course of five miles between Ft. Myer and Shuter's Hill near Alexandria. In this first cross-country flight the aeroplane flew at an average speed of 42.582 miles per hour. "The spectacle will probably never have its parallel in the history of our country," wrote James M. Cox of Dayton, later Governor of Ohio and Democratic nominee for President of the United States. "President Taft, Vice President Sherman and every member of the Cabinet, the Senate, the House, and the Supreme Court that was in Washington and able to be present, was there. Of course the public had made its contribution in numbers running into the tens of thousands. The ship in flight was noisy and Orville Wright sat out in front without any kind of protection against the elements. The course was to be from Ft. Myer to Alexandria and return. There was great excitement when the word was given and the propellers were turned over by hand and the machine catapulted over a track through the operation of a series of weights. The ship seemed to have the wings of a bird and rose proudly to a height of about three

Lt. Frank P. Lahm, the first rated military pilot, and Orville during the Ft. Myer trials of 1909— the 1909 Signal Corps Flyer, the world's oldest military aircraft, is on exhibit at the National Air and Space Museum, Washington, D.C.

Orville poses for the camera at Ft. Myer, 1909.

hundred feet. Soon we could see it circle the balloon. At that point Orville was compelled to lower his altitude in order to pass by the cable attached to the balloon. The ship passed from sight behind the hills on the horizon. Seconds grew into minutes and minutes seemed to be hours. The audience was in great suspense. Beads of sweat broke out over the forehead of Wilbur Wright. In our imaginations, anything could have happened. Finally the ship broke into the skyline and came rushing to the finish with a speed that seemed tremendous. The distance of ten miles was covered in fourteen minutes.''

The government gave the brothers a bonus of $5,000 representing ten percent for each mile over forty miles per hour. The machine was formally accepted August 2; the price was $39,000.

Orville and Katharine left Dayton August 8 for Berlin where Orville made several demonstration flights at Tempelhof before beginning to train flyers for the German Wright Company. His most distinguished passenger was the Crown Prince of Germany. Kaiser Wilhelm witnessed one of Orville's flights.

The Aero Club of America chose Glenn H. Curtiss to represent the country in an international aviation meet at Reims, August 22-29, 1909. Curtiss had won the *Scientific American* trophy for a fifty-two minute flight July 17 at Mineola, N.Y. Except the Wrights' flights, this was the longest made in the United States. It was made in an aeroplane built by the Aeronautic Society of New York, and it infringed the Wright patent on wing warping.

Orville to Wilbur

New York, August 10, 1909

I think best plan is to start suit against Curtiss, Aeronautic Society, etc., at once. This will call attention of public to fact that the machine is an infringement of ours. Mr. Cordley (Flint & Co.) favors starting suits in Europe at once and prosecuting them vigorously. If suit is brought before the races are run at Reims, the effect will be better than after.

Wilbur filed petitions for injunctions against Herring-Curtiss Company and Glenn H. Curtiss and against the Aeronautic Society of New York August 18, 1909.

1909 Signal Corps
Machine

This machine had a wingspan of 36 feet, 6 inches, a wing area of about 415 square feet, horizontal rudder area of about 80 square feet, and a vertical rudder area of about 16 square feet. It was 28 feet, 11 inches long and weighed 735 pounds.

Ernest L. Jones, editor of *Aeronautics*, decided to publish a statement on the Wright patent claim in his magazine. Chanute wrote to him, "I think the Wrights have made a blunder by bringing suit at this time. Not only will this antagonize very many persons but it may disclose some prior patents which will invalidate their more important claims." Chanute's letter to Jones was the beginning of an estrangement between Chanute and the Wrights.

While Orville was making demonstration flights in Germany, Wilbur made several flights in the United States. One flight, witnessed by hundreds of thousands of New Yorkers, was around the Statue of Liberty. He also flew from Governors Island up the Hudson beyond Grant's Tomb and back. He fastened a red canoe with ropes to the bottom of the machine for the twenty-one mile flight up the Hudson. No one had ever flown over the Hudson with the updrafts over the ships. Wilbur thought the canoe would serve as a float in case he came down.

He then left for College Park, Maryland, to fulfill the Army requirement of training flyers. His first student was Lieutenant Frank P. Lahm.

Now that the brothers had received both fame and money for their invention, they learned that others wanted a piece of the pie. George Spratt complained in a letter to them that he had made a suggestion to them about measuring lift and drift that they had used in a machine. He wanted credit for the suggestion, and said that the advice he had from them in return could not be considered in any way fair compensation.

His complaint disturbed Wilbur. He and Orville considered their own method of measuring lift and drift superior to Spratt's. "I suppose that when two men swap stories," he wrote in answer to Spratt's complaint, "each thinks his own better than the other's, and it is the same when men swap ideas. But aside from the ideas and suggestions you received from us, we also furnished copies of our tables, not only of the machine of which your idea formed a part, but also on the pressure-testing machine. My ideas of values may be wrong, but I cannot help feeling that in so doing we returned the loan with interest, and that the interest many times outweighed in value the loan itself."

He added that if Spratt felt he had not received enough value, he was ready to provide him with any scientific information or practical knowledge he had gathered in ten years of experimenting.

"Tell us your needs and we will help you," he said. "I learn from a gentleman I met in New York, who said he was associated with you, that you are experimenting with a machine on which you expect to mount a motor. Perhaps there is some point about which you would find our advice useful. I will be at College Park a few weeks and would be glad to have you come down."

Wilbur, Orville, and Katharine were all at home in early December. Wilbur said in a letter to Chanute that they were busy organizing their work and preparing their suit against Herring and Curtiss. "The affidavit filed by Mr. Herring is thoroughly characteristic of him. He has suddenly discovered that he invented in 1894 the method of controlling lateral balance by setting surfaces to different angles of incidence on the right and left sides of the machine and correcting the difference in their resistances by means of an adjustible vertical tail.

"We have closed out our American business to the Wright Company, of which the stockholders are Mssrs. C. Vanderbilt, Collier, Belmont, Alger, Berwind, Ryan, Gould, Shonts, Freedman, Nicoll & Plant. We received a very satisfactory cash payment, forty percent of the stock, and are to receive a royalty on every machine built, in addition. The general supervision of the business will be in our hands though a general manager will be secured to directly have charge. We will devote most of our time to experimental work."

Orville's Diary X, October 15, 1909: "Emperor come; Empress & Princess; Gen. von Plessen; Freiherr Holzing."

The Emperor was Kaiser Wilhelm II, the Empress was Augusta Viktoria, the Princess was Viktoria Luise, von Plessen was the Emperor's Aide-de-Camp, Holzing was Commandant of the Royal Stables. Bornstedt Field, near Potsdam, October 15, 1909

The first aeroplanes manufactured by the Wright Company were purchased by Robert J. Collier, Russell A. Alger, and Cornelius Vanderbilt. The price was $5,000 for an aeroplane.

While the Wright Company building was being erected, the brothers rented a room in the Speedwell Motor Car plant. Gilbert J. Loomis, designer of the Speedwell, in later years said that when agents of the company came to Dayton, the first thing they wanted to do was to go out to Simms Station to see the brothers fly. "The Wright brothers, despite the fact that they would go up in a glorified kite, were very timid. We always had several cars going out to the flying field, but although we offered them a chance to ride in one of them, they said 'Automobiles are too dangerous,' and they preferred to hang on to the running board of an overloaded traction car. Wilbur remarked to me one day that there ought to be a law passed prohibiting the manufacture of cars that could go more than twenty miles an hour."

Arnold Kruckman, aeronautical editor of the New York *World*, said of the Wrights in an article December 12, "Their persistent failure to acknowledge their monumental indebtedness to the man who gave them priceless assistance has been one of the puzzling mysteries of their career." Wilbur answered the attack in a candid letter, not only praising Chanute for his long friendship and constant encouragement, but also pointing out that most of information that leaked out about the Wrights' work had come from Chanute, a fact that led to the false impression that they were working under his direction and with his financial assistance. "Many of the published stories have been very embarrassing because if left uncorrected they tend to build up a legend which takes the place of truth, while on the other hand, any attempt on our part to correct inaccuracies gives us the appearance of ungratefully attempting to hurt the fame of Mr. Chanute. Rather than subject ourselves to criticism on that score we have preferred to remain silent, but now you fault with our silence. We, rather than Mr. Chanute, have been the sufferer from this silence so far.

"Chanute is one of the truest gentlemen we have ever known and a sympathetic friend of all who have the cause of human flight at heart. But we hope that you will make certain of your facts before making an accusation again."

Kruckman quoted Chanute as saying that the Wrights' claim to have been the first to maintain lateral balance by adjusting the wing tips to different angles of incidence could not be maintained, as the idea was well known already when they began their experiments.

"As this opinion is quite different from that which you expressed in 1901 when you became acquainted with our methods," Wilbur wrote January 20, 1910, "I do not know whether it is mere newspaper talk or whether it really represents your present views. So far as we are aware the originality of this system of control with us was universally conceded when our machine was first made known, and the questioning of it is a matter of recent growth springing from a desire to escape the legal consequences of awarding it to us." He told Chanute if there was anything in print that invalidated their rights, he would like to know it before they spent a great amount of money on lawyers.

Chanute answered that he did tell them in 1901 that the mechanism by which their surfaces were warped was original with the Wrights. But, he went on, "this does not follow that it covers the general principle of warping or twisting wings, the proposals for doing this being ancient." He mentioned five earlier experimenters, including Mouillard. Chanute had given them a copy of Mouillard's patent of 1901 and mentioned his method of twisting the rear wings. "If the courts will decide that the purpose and results were entirely different and that you were the first to conceive the twisting of the wings, so much the better for you, but my judgment is that you will be restricted to the particular method by which you do it. Therefore it was that I told you in New York that you were making a mistake by abstaining from prize-winning contests while public curiosity is so keen, and by bringing suits to prevent others from doing so. This is still my opinion and I am afraid, my friend, that your usually sound judgment has been warped by the desire for great wealth."

Chanute also said that he inferred that his beliefs were a grievance in Wilbur's mind and that he himself had a little grievance against Wilbur. "In your speech at the Boston dinner, January 12th, you began by saying that I 'turned up' at your shop in Dayton in 1901 and that you then invited me to your camp. This conveyed the impression that I had thrust myself upon you at that time and it omitted to state that you were the first to write me, in 1900, asking for information which was gladly furnished, that many letters passed between us, and that both in 1900 and 1901 you had written me to invite me to visit you, before I 'turned up' in 1901. This, coming subsequently to some somewhat disparaging remarks concerning the helpfulness I may have been to you, attributed to you by a number of French papers, which I, of course, disregarded as newspaper talk, has grated upon me ever since that dinner, and I hope, that in future, you will not give out the impression that I was the first to seek your acquaintance, or pay me left-handed compliments, such as saying that sometimes an experienced person's advice was of great value to young men."

The brothers found the letter incredible. In a lengthy reply on January 29, Wilbur took up the charges Chanute had made. "It came to us as a shock," he said, "when you calmly announced that this system was already a feature of the art well known, and that you meant only the mechanical details when you referred to its novelty. If the idea was really old in the art, is it somewhat remarkable that a system so important that individual ownership of it is considered to threaten strangulation of the art was not considered worth mentioning then, nor embodied in any machine built prior to ours." He took up the five earlier experimenters Chanute had mentioned and ticked them off, one by one, saying that he found in none of their writings any mention whatever of controlling lateral balance by adjustments of wings to different angles of incidence on the right and left sides. "Have you ever found any such mention?" he asked Chanute. "Unless something as yet unknown to anybody is brought to light to prove the invention technically known to everybody prior to 1900, our warped judgment will probably be confirmed by the other judges as it was by Judge Hazel at Buffalo. (Judge John R. Hazel of the Federal Circuit Court of Buffalo, New York, had granted on January 3,

1910, to the Wright Company a preliminary injunction restraining the Herring-Curtiss Company and Glenn H. Curtiss from manufacturing, selling, and using the Curtiss aeroplane for exhibition purposes.)

"As to the inordinate desire for wealth," Wilbur continued, "you are the only person acquainted with us who has ever made such an accusation. We believed that the physical and financial risks which we took, and the value of service to the world, justified sufficient compensation to enable us to live modestly with enough surplus income to permit the devotion of our future time to scientific experimenting instead of business. We spent several years of valuable time trying to work out plans which would have made us independent without hampering the invention by the commercial exploitation of the patents. These efforts would have succeeded but for jealousy and envy. It was only when we found the sale of the patents offered the only way to obtain compensation for our labors of 1900-1906 that we finally permitted the chance of making the invention free to the world to pass from our hands. You apparently concede to us no right to compensation for the solution of a problem ages old except such as is granted to persons who had no part in producing the invention. That is to say, we may compete with mountebanks for a chance to earn money in the mountebank business, but are entitled to nothing whatever for our past work as inventors. If holding a different view constitutes us almost criminals, as some seem to think, we are not ashamed. We honestly think that our work of 1900-1906 has been and will be of value to the world, and that the world owes us something as inventors, regardless of whether we personally made Roman holidays for accident-loving crowds."

He said he could not say without seeing the French papers in which Chanute said he had read disparaging remarks whether they quoted him accurately or not. He said that he had heard in France that he and Orville had taken up aeronautical studies at Chanute's special instigation, that they had obtained their first experience on one of his machines, that Chanute had provided money, and that "in short, that you had furnished the science and money while we contributed a little mechanical skill, and that when success had been achieved you magnanimously stepped aside and permitted us to enjoy the rewards."

Wilbur said that he had never spoken out because he did not wish to disparage Chanute or hurt his feelings. "I cannot understand your objection to what I said at the Boston dinner about your visit to Dayton in 1901. I certainly never had a thought of intimating either that you had or had not been the first to seek an acquaintance between us. You also object to my expressing an appreciation of the influence which your friendship had on our work and lives. One of the *World* articles said that you had felt hurt because we had been silent regarding our indebtedness to you. I confess that I have found it most difficult to formulate a precise statement of what you contributed to our success. We on our part have been much hurt by your apparent backwardness in correcting mistaken impressions, but we have assumed that you too have found it difficult to substitute for the erroneous reports a really precise statement of the truth. If such a statement could be prepared it would relieve a situation very painful to both you and to us.

"I have written with great frankness," he said in conclusion, "because I feel that such frankness is really more healthful to friendship than the secretly nursed bitterness which has been allowed to grow for so long a time. I expect that we will always continue to disagree in many of our opinions just as we have done ever since our first acquaintance began and even before, but such differences do not need to disturb a friendship which has existed so long."

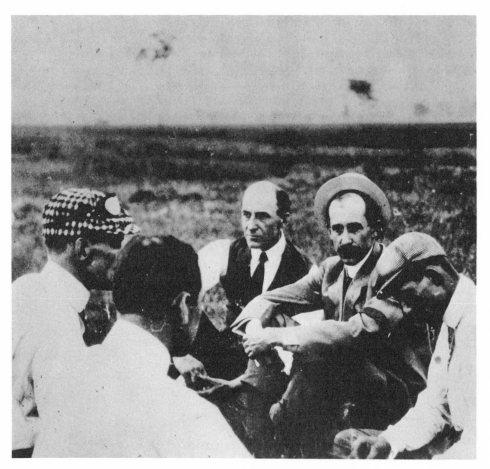

Instructors Wilbur and Orville chat with pupils Frank Coffyn, with goggles, Ralph Johnstone, and Walter Brookins at far right. In 1910 Brookins flew a record 192½ miles from Chicago to Springfield, Ill. Wilbur followed in a special car attached to an Illinois Central train.

After he received Wilbur's July 19 letter, Chanute wrote about the quarrel to George Spratt. "I will answer him in a few days," Chanute said, "but the prospects are that we will have a row. I am reluctant to engage in this, but I think I am entitled to some consideration for such aid as I may have furnished."

Nor was Chanute the only person other than Spratt who did not rejoice in the Wrights' success. Dr. Albert F. Zahm, whom they had long known, wrote asking to be employed as a witness for the Wrights in the Curtiss suit. When the Wrights did not employ him, he became a witness for Curtiss instead.

On February 8, 1910, the brothers went to Washington to accept the first Langley Medal "for especially meritorious investigation in connection with the science of aerodromics and its application to aviation." The medal, a disk about three inches in diameter shows a woman in flowing draperies sitting on top of the world, holding in her left hand a torch and in her right, a scroll. Supreme Court Justice Melville W. Fuller presented the medals, one to each brother, February 10. The Washington *Herald* of that week said, "The scene was almost the picture of school boys approaching at the end of term, to receive the award of well-doing from the headmaster."

During the spring of 1910 Orville trained pilots for the Army at Montgomery, Alabama. The first pilot he trained was a Dayton civilian, Walter Brookins, who then stayed behind to train other pilots for the Army when Orville returned to Dayton to open a flying school.

Brookins joined Orville at Simms Station, and they jointly trained a group of pilots which formed the nucleus of instructors who along with military men, trained tens of thousands of pilots.

When a student signed up for flying lessons at the Wright factory and paid $60 in advance, he was requested to spend time at the factory practicing control movements on a trainer set up in one section of the plant. The trainer—in effect the first Link trainer—was a Wright Model B without the tail section and engine. It was cradled so that it could oscillate laterally an electric motor driving a cam, which continuously changed the pattern of lateral movement. When the student moved the combination warp and rudder lever correctly, the attitude of the wings would be brought back to a level position. After a few hours' practice, the movement to correct imbalance became instinctive.

Then the student reported to his instructor at Simms Station. The instructor explained the instrumentation, which consisted of a piece of string about twelve to fourteen inches long tied between two collars soldered to the wire stretched between the forward ends of the skids. The string served a useful purpose; if turns were being properly executed, it trailed straight back. If the machine skidded or slipped, the string trailed off to one side or the other. Now the student was ready for his first flight.

Both instructor and student were dressed for flying when they turned their caps around backwards. If the student collided with a bug in the air, he wore a pair of automobile goggles. The student gradually took over the controls, first on straight-aways, then on right and left turns, then figure eights, and landings. When the student was able to handle the machine for a full circuit of the field the instructor sat with his arms folded. When the student encountered a patch of turbulent air, the instructor turned his head toward him with a broad grin on his face but he did not touch the controls.

The Model B was a biplane with a wingspan of about 38 feet and a chord of about six feet. Total wing area was 150 square feet. It was about 30 feet long and stood eight feet high. It weighed loaded around 1,250 pounds. It had a 35-horsepower engine, which turned a maximum of 1,400 rpm, driving two pusher propellers geared down to turn about 450 rpm.

The landing gear consisted of two skids attached by struts to the central portion of the wing structure extending well forward and two or three feet behind the lower wing. Two rubber-tired wheels straddled each skid. The weight of the plane at rest was supported by the wheels and the rear portion of the skids.

The instructor and student sat side by side on seats on the lower wing, their feet resting on the light structure forward. The engine was on the wing to right of center, the gas tank above and rearward, and a long narrow radiator was mounted vertically between the wings.

The machine had two complete sets of controls for dual instruction, each set having two levels, one of which had two functions. The left-hand lever in each set was moved fore and aft, a normally instinctive motion, to ascend or descend. The right-hand or wing warp lever actually consisted of two separate levers in one. To make a left turn, this lever was moved forward, causing the machine to bank to the left. At the same time the left rudder was automatically applied through a linkage.

The Model B had a top speed of 50 miles an hour and a landing speed of 25 to 30 miles an hour.

To take advantage of the smooth early and late evening air the students often slept on cots in the machine shed, taking meals at a nearby farmhouse. When the weather was bad, the students stayed at the factory. Orville Wright always was willing to talk to the students, but while he carried on a conversation, he picked up the brass screws that had been brushed off the work benches during the day. He had learned that each item in the factory was of intrinsic value and not to be wasted.

Wilbur and Orville, after receiving the Langley Medal February 10, 1910, leave the Smithsonian Institution with Dr. Charles Walcott and Alexander Graham Bell.

When the instructor decided that his student was ready to fly alone, he said, "Now you fly two or three circles around the field and land. Don't try anything else."

The Fédération Aéronautique Internationale, organized in 1905 to keep records of flights of aeroplanes and balloons, issued licenses. It was represented in the United States by the Aero Club of America. Three observers witnessed the tests which involved climbing to an altitude of more than 500 feet, doing a series of figure eights around two pylons, and landing within 350 feet of a designated point. One of the pylons at Simms Station was a pole with a flag and the other was an old tree.

The entire Wright family rode on the traction to Simms Station on May 25. Orville made a flight with Wilbur as passenger, the only time the two ever flew together.

On the same day Orville took eighty-two-year-old Bishop Milton Wright for a flight. In his diary the bishop noted, "Orville took me up 350 feet and 6.55 minutes."

Chanute did not answer Wilbur's letter of January 29, 1910, and on April 28 Wilbur wrote again, trying to soothe Chanute's hurt feelings and again suggesting a joint statement. Chanute replied in May that he had not been feeling well and was about to leave for Europe. He said that he had paid no attention to the newspaper stories "being aware how newspaper reporters made mountains out of molehills, yet the question arose in my mind whether there had been a molehill. I am sure in my mind that in my own case there has not been a molehill of disparagement of your achievements or claim that I was entitled to part of the credit.

"The difference between us, i.e., whether the warping of the wings was in the nature of a discovery by yourselves, or already had been proposed and experimented by others, will have to be passed upon by others, but I have always said that you are entitled to immense credit for devising apparatus by which it has been reduced to successful practice.

"I hope, upon my return from Europe, that we will be able to resume our former relations."

It was the last letter in the long correspondence between Octave Chanute and the Wright brothers.

In October of 1910 the Wright Company workers completed the *Baby Grand*, the new Wright Model R, powered by an eight-cylinder engine that attained speeds of between 70 and 80 miles an hour. They planned to compete with it in the Gordon Bennett International Aviation Cup contest at Belmont Park, New York. The meet opened October 22 and two days later in practice Orville flew almost 70 miles per hour. In the race on October 29, however, Walter Brookins smashed the machine on take-off. The winning speed by Claude Grahame-White was 61 miles per hour.

Orville sailed from New York September 15 to inspect the European business. He found that the business in France and in Germany had deteriorated and wrote to Wilbur, "I have about made up my mind to let the European business go. I don't propose to be bothered with it all my life and I see no prospect of its ever amounting to anything unless we send a representative to watch our interests."

Wilbur was more optimistic. He pointed out that they might expect that in their absence the French and German companies would do foolish things, but he advised insistence that all matters be rectified and assurances for the future made. "I notice that an effort is being made to erect a monument to Lilienthal and to provide something for his family. If you find the business is in proper hands and their plans sensible and reasonable, and the need of this family real, we might contribute anything you think best up to a thousand dollars." Orville called on Mrs. Lilienthal and made the cash contribution Wilbur had suggested.

At home the business had its ups and downs. With Roy Knabenshue as chief promoter, the Wrights had organized an exhibition company. Mabel Beck, a Daytonian, became Roy's secretary and remained with the Wrights after the exhibition company was disbanded. Although the company had some trouble in collecting for exhibitions, still the first year proved profitable with $100,000 net profit. The average charge for exhibition flights was $5,000.

Octave Chanute died November 23, 1910. He was almost seventy-nine years old. Wilbur wrote to George Spratt a month later, "You know, of course, of the death of Mr. Chanute. Orville and I were at a dinner in his honor at Boston in January, at which time he seemed in pretty good health. But in May he wrote that he was intending to go abroad for his health. He had a severe attack of pneumonia while in Germany and France, which left him so feeble that he had scarcely strength to get home. There he gradually became weaker and faded away. I attended his funeral at his home on Dearborn St. Only a few invited friends were present, mostly his associates in the Society of Engineers. Colonel William A. Glassford of the Signal Corps was the only one of his aeronautical friends I recognized. He was buried at Peoria beside his wife."

Wilbur wrote an article summing up Chanute's life work in the January 1911 *Aeronautics,* concluding, "His writings were so lucid as to provide an intelligent understanding of the nature of the problems of flight to a vast number of persons who would probably never have given the matter study otherwise, and not only by published articles, but by personal correspondence and visitation, he inspired and encouraged to the limits of his ability all who were devoted to the work. His private correspondence with experimenters in all parts of the world was of great volume. No one was too humble to receive a share of his time. In patience and goodness of heart he has rarely been surpassed. Few men were more universally respected and loved."

Orville pilots the modified 1911 glider on one of the final flights at Kitty Hawk. He sets a world soaring record of 9 minutes, 45 seconds.

11

Beyond Return

*For we needs must die, and are as water spilt upon the ground,
which cannot be gathered up again.*

—2 Samuel 14:14

The National Geographic Society invited Wilbur to attend its sixth annual banquet in January 1911, a dinner in honor of the United States Army and of the discovery of the art of aviation.

Speakers were Henry Gannett, president of the Society; General John M. Wilson; German Ambassador Count J. H. von Bernstorff; Mexican Ambassador Francisco Léon de la Barra; British Ambassador James Bryce; President William Howard Taft; Chief of Staff General Leonard Wood; General Adolphus W. Greely, former chief of the Signal Corps, and Wilbur Wright.

Wilbur Wright to the National Geographic Society

Washington, January 14, 1911

Mr. Chairman, Ladies and Gentlemen: When the Director of your Society extended me the invitation to be present at this dinner he indicated that a few remarks might be required of me. This game of after-dinner speaking is one that I have never played very much; so I said to myself, "It will be a good thing to study up on the rules of the game."

So last week I went up to New York and had an opportunity to attend a dinner there at which were present a number of distinguished speakers. The first gentleman on the program was a bishop, and he made a splendid speech. I said to myself, "If it is necessary for me to speak like that I might as well give up trying to play this game." But as the evening wore on I began to discover things. I found out after a little bit that this man had been stealing the speeches of all the men who were to follow him on the program; and not only that, but according to the statements of the other speakers, he had not only stolen their ideas, but he expressed them better than they could express them themselves. "Well," said I, "if a man is permitted to steal the speeches of all the other men on the program, it is an easy job." So when I heard that among the speakers this evening were some of our most distinguished men, including the President of the United States, I said to myself, "This will be an easy job. I will steal their speeches and I will present them better than they could present them themselves."

But unfortunately the novice often learns some other rule that is equally as important. In this case the rule I overlooked was the rule that provides an arrangement of the speakers on the program. Instead of being on first, so that I could steal their speeches, I have been put on last, with the result that all the speech I was going to make they have made; and, as usual in banquets, according to rule, they have said it a great deal better than I could say it myself.

But I have discovered one other rule that is customary in this game, and that is to choose the subject which is appropriate to the occasion; and another rule: confine your remarks strictly to the subject. Therefore I chose for my subject this evening "geography." I shall confine my remarks more particularly to that branch of geography which I discovered on the front of a book once in my father's library. The book was entitled "The Geography of the Heavens." I am not certain now whether the book treated of astronomy or whether it was a book on theology. For my part I steered clear both of astronomy and of theology; and, in order to keep off the ground that might have been covered in that book, it is my intention to turn the subject exactly upside down, and instead of viewing it from the ground I will view geography from the heavens.

The ways and means of studying the earth from above have now become pretty well known. Some of the uses which may be expected to be derived from the aeroplane and similar instruments have been told you very well by our distinguished friend, General Wood. The real uses of the aeroplane in warfare are so much better known to him and have been so much better presented that I shall not attempt to take up your time with that. The advantages of knowing what the enemy is doing, with the consequent advantage of being able to concentrate your own troops at the critical spot—the advantage of rising high in the air for the purpose of determining the accuracy of gun power and giving appropriate directions for carrying on work of that kind—are so well known that it is useless for me to take it up now.

The leading nations of the earth are taking up the subject, our own nation being the first of all to begin it. But, unfortunately, there seems to be some hesitation at present. I do not know exactly what the trouble is. I presume possibly that if we were to apply the parable of the talents we would possibly arrive at something near the present situation. The Department probably feels that with the small equipment it now has it is useless to do anything. On the other hand, Congress seems to feel that unless something is done with the equipment already on hand it is useless to appropriate for it now. I hope the day will soon come when more will be done with what we already have, so that Congress will see fit to appropriate more if it sees that the money will be well spent. I thank you very much.

Much of the brothers' energy in 1911 was spent on the lawsuits in the courts, on flying exhibitions, and on the training of flyers. Henry Peartree, the lawyer for Flint and Company, telegraphed that the Wrights should come to Paris to testify in a French patent case. Wilbur left Dayton March 12 and did not return till August. While he was gone, the United States Navy ordered a Wright plane for $5,000, and Lieutenant John Rodgers of the Navy came to Dayton for flying lessons. Lieutenants Henry (Hap) Arnold and Thomas Milling of the Army also came to Dayton for lessons.

Wilbur to Orville

Berlin, June 28, 1911

The men who had almost finished their training when I arrived in Berlin ten weeks ago are still almost trained but cannot fly alone yet! The poor Captain cannot understand how you can train men in a week at home. He would not believe that I could carry two men with 375 turns of the propellers till I took him up and did it.

Orville met Wilbur in New York on his return from Europe August 9. The two attended an executive meeting of the Wright Company and returned to Dayton. Orville left the next day for Chicago for the Chicago International Aviation Meet in which the Wright Company had seven entrants.

While Wilbur was testifying in patent suits, Orville went to Kitty Hawk October 7 to try out a new stabilizer, but the presence of so many reporters prevented him from using it. Orville made about 90 glides from Kill Devil Hill, West Hill, and Little Hill, and October 24 he set a new soaring record of nine minutes, 45 seconds, in a 50-mph wind. This record stood for ten years.

Wilbur and Orville on November 4, 8, and 9 experimented flying a Curtiss machine at Simms Station to become familiar with the control mechanisms which they claimed infringed the Wright patents, then in suit in the courts in New York.

Judge Learned Hand, of the United States Circuit Court, granted an injunction on December 12, 1911, to the Wright Company to restrain English flyer Claude Grahame-White from flying in the United States without permission of the Wrights. The next day the Wright Company sued Grahame-White for $50,000 to obtain an accounting of profits from flights made by him in the United States before November 29, 1910.

Wilbur to M. Hévésy

Dayton, January 25, 1912

During the past three months most of my time has been taken up with lawsuits and I have been away from home most of the time. I am hoping to be freed from this kind of work before another year has ended. It is much more pleasant to go to Kitty Hawk for experiments than to worry over lawsuits. We had hoped in 1906 to sell our invention to governments for enough money to satisfy our needs and then devote our time to science, but the jealousy of certain persons blocked this plan, and compelled us to rely on our patents and commercial exploitation. We wished to be free from business cares so that we could give all our time to advancing the science and art of aviation, but we have been compelled to spend our time on business matters instead during the past five years. When we think what we might have accomplished if we had been able to devote this time to experiment, we feel very sad, but it is always easier to deal with things than with men, and no one can direct his life entirely as he would choose. Yet these years have not been without their pleasant spots, and we look back with much enjoyment to the friendships made during this period.

If you should come to America do not forget Dayton.

Wilbur testified from February 2 to March 14, 1912, in Dayton in *Wright Company v. Herring-Curtiss Company and Glenn Curtiss* patent suit.

Wilbur to E. C. Huffaker

Dayton, April 6, 1912

I was very glad to receive your letter of last December, and to know that you are alive and well. At that time we were thinking of calling you as a witness in one of our numerous patent suits, but the amount of other evidence was becoming so big that we finally decided that it was not necessary.

It is rather amusing, after having been called fools and fakers for six or eight years, to find now that people knew exactly how to fly all the time. People who had not the least idea of flying until within the last year or two now attempt to write books stating what the situation of the flying problem was in 1900 and 1901, when we made our first experiments at Kitty Hawk. In view of our experiences in 1901 it is amusing to hear them tell that the science of aerodynamics had been reduced to a very exact basis, so that anyone could calculate without difficulty the lift and drag of aeroplane surfaces.

Wilbur and Orville were working on one of their Flyers at Simms Station on Thursday, May 2, 1912. "Orv, I don't feel well."

Orville glanced up from his work. "What's wrong?"

"I have a headache and a chill and feel weak."

"Maybe you're tired from your trip. You think we should go home?"

Wilbur shook his head. "Perhaps it's something I ate in Boston. I had oysters and clams there yesterday and this isn't a month with an *r* in it. No, let's finish our work for the day."

The following day Wilbur felt worse, but he spent the day working. On Saturday, May 4, he stayed in bed and asked Orville to go for Dr. Spitler. Dr. Spitler was out of town, but his partner Dr. Daniel Conklin came. After examining Wilbur, he said, "His temperature is 102. His cough could indicate pneumonia, or there is a possibility of malaria. Give him cool baths to try to get the temperature down."

As the days passed, Wilbur's temperature continued to fluctuate between 103° and 105° and his chief complaint was a severe headache.

On May 11 Dr. Levi Spitler asked Wilbur whether he had eaten shellfish within the past two weeks. Wilbur answered, "I did in Boston."

"I'll be back this afternoon." Dr. Spitler talked downstairs with the family. "Wilbur has abdominal distension and colicky pain. The temperature remains high. Because Wilbur ate shellfish, I suspect typhoid. You probably know that typhoid epidemics can result from eating shellfish caught in areas where sewage is dumped. Wilbur must have nursing care and I will send for Mrs. Sullivan. You must all take precautions when you visit him. Do not touch the bedclothes, and when you leave the room, wash your hands with soap and water."

Wilbur's temperature continued to range between 103° and 105°. On May 18 he suffered delirium and Dr. Spitler administered opiates.

When people in Dayton and other parts of the world learned of Wilbur's illness, so many telephone inquiries came to 7 Hawthorn Street that the family ordered the Bell exchange to limit the calls. "For two weeks the pretty elm and maple studded little thoroughfare was in the throes of an awful suspense," wrote a reporter for the *Dayton Evening Herald.* "Not a piano or other musical instrument could be heard for many rods because all realized the gravity of the situation and all the neighbors wished to contribute not the slightest disturbance to the ease and comfort of their beloved friend and neighbor. Even the little children played, not in front of their homes but conscious of the situation, ambled to other streets for indulgence in their miniature armies, their doll house construction, their hobby horse rides, their sports and games. With spontaneity these little folks seemed to act conscious, fully conscious, of the impending danger. Even grown-ups tiptoed past the modest white painted Wright home and a singular stillness pervaded the atmosphere save for the ring, now here, now there, of a telephone bell in its mission of inquiry."

Bishop Wright's Diary

May 19, 1912

Wilbur ceases to take opiates but is mostly quiet and unconscious. His sickness is very serious. He is mostly unconscious.

May 20, 1912

Dr. Spitler came afternoon and at night with Dr. D. B. Conklin. Wilbur's case very serious. He notices little.

May 22, 1912

The doctors had Dr. Busheimer of Medical College Cincinnati come in to consult in Wilbur's case, but could suggest no further treatment. The doctors think him better.

May 26, 1912

Wilbur was worse in the night. Orville slept but little. Katharine received a telegram from Lord Northcliffe. Many cards and telegrams.

May 27, 1912

Both Conklin and Spitler came at 7:00 morning. They think the case very bad. His temperature was higher and he has difficulty with the bladder, and his digestion is inadequate. Reuchlin (from Kansas) saw him in the afternoon. I slept with my clothes on. We thought him near death. He lived through the evening.

May 28, 1912

I slept some in the night. I awoke at 4:00. Wilbur is sinking. The doctors have no hope of his recovery. Mr. Toulmin called. At 6:30 eve the doctor thought him dying.

May 29, 1912

Wilbur seemed no worse, though he had a chill. The fever is down, but rose high. He remained the same till 3:15 in the morning, when, eating his allowance 15 minutes before his death, he expired, without a struggle. His life was one of toil. His brain ceased not its activity till two weeks of his last sickness had expired. Then it ceased.

May 30, 1912

This morning at 3:15, Wilbur passed away, aged 45 years, 1 month, and 14 days. A short life, full of consequences. An unfailing intellect, imperturbable temper, great self-reliance and as great modesty, seeing the right clearly, pursuing it steadily he lived and died. Many called—many telegrams. (Probably over a thousand.)

May 31, 1912

Boyers are the undertakers. We get many letters and telegrams and cablegrams of sympathy from all people of every sort, and from all societies, and from dignitaries. Flowers come from individuals and societies, most beautiful. Arrange with undertaker.

June 1, 1912

I awakened before midnight and got up near two hours. I slept then till nearly 5:00. Arose and washed off and dressed till near 7:00. The undertakers put Wilbur in the burial casket. Took him to the church at nearly ten. Many relatives come; many friends. Wilbur's body lay in state at First Presbyterian Church from 10 till 1:00. Rev. Maurice Wilson assisted by J. Howe conducted funeral services 3:00.

June 3, 1912

Wilbur is dead and buried! We are all stricken. It does not seem possible that he is gone. Probably Orville and Katharine felt his loss most.

Dayton Evening Herald, May 30, 1912

FIGHTING GRIM DEATH
TO LAST, BRAVE AIRMAN
GENTLY FALLS ASLEEP

Marvelous Courage, Tenacity and Vitality
Displayed by Aviator Astound
Physicians and Attendants.

PROFOUND SORROW IS FELT
IN ALL SECTIONS OF CITY

His Achievements Have Brought Honor
to Himself, His Nation, His
Family and Dayton.

With the same heroism that characterized his life and his struggles to overcome the problem of air flight, Wilbur Wright's contest with the unconquerable spirit of death was terminated at 3:15 Thursday morning.

DECREE OF PROVIDENCE.

By a singular degree of Providence, Memorial Day, established as a national day of tribute to the nation's martyred dead, became a day of national, even international, tribute to a martyr who sacrificed energy, strength, skill—his very life— to the advancement of science and the improvement of the conditions of mankind.

DEATH COMES QUIETLY.

Dissolution came quietly, without a struggle, in the still hour of the early morning with the distinguished patient surrounded by his venerable father, Bishop Milton Wright; his equally famous brother, Orville; his devoted sister, Katharine; the other two brothers, Lorin and Reuchlin, and Dr. Daniel Beckel Conklin, who has been in constant attendance since he took to bed on Saturday, May 4.

Among the tributes that came to 7 Hawthorn Street were messages from President William Howard Taft, General Leonard Wood, General James Allen, Secretary of War Henry L. Stimson, and scores of high-ranking Army officers. Taft's message read, "I am sorry that the father of the great new science of aeronautics is dead and that he has not been permitted to see the wonderful development that is sure to follow along the primary lines which he laid down. He deserves to stand with Fulton, Stevenson and Bell."

One of the telegrams of condolence came from Glenn H. Curtiss. All public offices and stores in Dayton closed during the funeral, and street railway cars halted for three minutes from 3:30 to 3:33 p.m. Among the honorary pallbearers were Robert J. Collier, John H. Patterson, the Honorable James M. Cox, Dr. Levi Spitler, and Dr. D. B. Conklin. Wilbur Wright was buried in Woodland Cemetery in Dayton.

In his will, written May 10, 1912, Wilbur left a few small bequests and $50,000 each to Reuchlin, Lorin, and Katherine. "I hereby give to my father, Milton Wright of Dayton, Ohio," the will said, "my earnest thanks for his example of a courageous, upright life and for his earnest sympathy with everything tending to my true welfare, and in addition I give and bequeath to him the sum of One Thousand Dollars (1,000) which I desire him to use for little unusual expenditures as might add to his comfort and pleasure."

The entire remainder of his estate he bequeathed to Orville, "who has been associated with me in all the hope and labors both of childhood and manhood, and who, I am sure, will use the property in very much the same manner as we would use it together in case we would both survive until old age. And for this reason I make no specific bequest to charity."

The Aeronautical Society of Great Britain opened a subscription fund to establish a memorial to Wilbur in appreciation "of his great work and in recognition of the support he gave to the Aeronautical Society of Great Britain." It was to be an annual lecture in Wilbur's honor. Griffith Brewer was in charge of the organization and collection of the fund, and he gave the first Wilbur Wright Memorial Lecture. In France Baron Paul d'Estournelles de Constant began a subscription for the building of a monument at the Auvours training camp at Le Mans.

Bishop Wright, who was eighty-four years old, wrote in his diary, "In memory and intellect, there was none like him; he systemized every thing. His wit was quick and keen. He could say or write anything he wanted to. He was not very talkative. His temper could hardly be stirred. He wrote much. He could deliver a fine speech, but was modest."

When a Dayton *Journal* reporter asked Orville for a statement, he said, "The death of Brother Wilbur has been an irreparable blow to all of us. It comes as a fearful shock right when we were in the midst of plans for a future bright with promise.

"How it will affect our organization at this time I cannot say. I have given no thought to the business affairs since his illness developed the fatal turn.

"Of course I will continue my work at the factory but I do not know how his place will be filled. It will be the most difficult task we have yet faced."

"In this picture at Kitty Hawk in 1911 I am the ten-year-old boy sitting on the sand. Uncle Orv saw some newspapermen hiding in the sand hills, and he invited them to come to our camp. They are the men in this picture.

"Of all the family, Uncle Orv took Uncle Will's death the hardest. They were so close to each other. They had done things since they were kids. To tell the truth, Uncle Orv never got over Uncle Will's death.

"Uncle Will liked kids who asked questions, and I was always asking him questions. That's why he liked me."
—Horace Wright, Dayton, 1983

12

Pilots of the Purple Twilight

I believe the day is at hand when the flyer will be entirely relieved of the work of maintaining the equilibrium of his machine and that his attention will be required only to keeping it on its course and in bringing it safely in contact when landing.

—Orville Wright, 1914

Orville to Reuchlin

Dayton, August 16, 1912

I think it was Will's wish to have all the Wright Company interest stay in my hands. I am sure it is the desire of the other stockholders. The stock is now in the hands of only a few persons, so that it makes it very much easier to transact the business. We occasionally wish to do things that require the consent of each stockholder, and, of course, the more the stock is distributed, the harder it is and the more time it takes to get such consent. In drawing up his will, I think Wilbur thought that we had a little over $300,000 besides our Wright Company stock. Our profits of last year added to what we had the year before would have made our total above $300,000, but together we gave away a little over $20,000 last year, so that this reduced his share about $10,000. I do not think there is any question about our winning our patent suits, but of course there is no certainty in the law; but in case we should lose the suits, and in case no value were placed on the patents, Will's stock in the company ought to have an actual value of about $25,000.

It is very generous of you to offer to take only ¼ of the property Will possessed outside of his Wright Company stock, but I am glad to have you get the full amount as he left it. The stock can hardly help but have some value, and even if it had none, I am fixed so that I will get along very comfortably.

During late 1912 and in 1913 Orville worked both on his automatic stabilizer and an aeroplane equipped with pontoons. He made a trip to Germany in February 1913 to testify in patent suits. The German Supreme Court at Leipzig rendered a decision favorable to the Wrights February 26.

In the United States February 27 Judge John R. Hazel, U.S. District Court in Buffalo, granted the Wright brothers' petition for an order restraining Glenn H. Curtiss and others from the manufacture and sale of infringing wing-warping machines.

Judge Hazel said in part, "I am of the opinion, after complete consideration of the testimony on both sides, that the patentees by their method of securing equilibrium of the aeroplane made an important advance in an embryonic art. They were not the first to conceive the idea of using monoplane or biplane surfaces for flying, nor the first to support two planes at their margins one above the other, or to use vertical tails, or rudders for steering, or to place horizontal rudders forward of the machine to guide it upward or downward in its flight. The prior separate use of such elements is freely admitted by the patentees, but they assert, rightly, I think, that the patented combination was a new combination performing a new and novel result. Having attained success where others failed, they may rightly be considered pioneer inventors in the aeroplane art. The employment, in a changed form, of the warping feature or its equivalent by another, although better effects or results are obtained, does not avoid infringement.

"The witness, Curtiss, frankly testified that the purpose (of the ailerons) is to preserve the lateral balance 'without the use of any other element or part,' making no difference whether the aeroplane is in a straight or curved flight. Such concession supports the infringement of the claim under consideration."

Curtiss appealed the decision. While his appeal was being considered in the U.S. Circuit Court of Appeals, Second Circuit, New York, Curtiss continued operations by posting bond. At this time—1913—the Regents of the Smithsonian awarded him the Langley Medal, the first time it had been given since the Wright brothers received the medals in 1910. Curtiss received the medal for the development of his flying boat.

While Orville and Katharine were sailing to New York on March 13, the French courts rendered a patent decision in favor of the Wrights. They returned to a rain-soaked Dayton March 19. Heavy rains increased on Easter Sunday, March 23, and continued all day and night. By Monday small towns to the north of the city were under water. On Tuesday morning the levees along the banks of four streams that join in the city allowed rivers carrying two and one-half times their normal amount of water to flow into the city.

Water poured into Hawthorn Street about 8:00 a.m. March 25. It quickly filled the Wright basement and began to rise to the first floor. "I put on my overcoat ready to go," the bishop wrote in his diary. "A canoe came for Mrs. E. Wagner, and the boys said I could get in, too. It glided down Hawthorn and on Williams Street to William Hartzell's next north of the Baptist Church. I walked in the dooryard—saved my shoes." Orville and Katharine were still sleeping when their father left. They had half an hour to get out of the house. A moving van took them to the home of their friends, the E. S. Lorenzes. Because she did not know where her father was, Katharine put up a sign asking anyone who knew where he was to get in touch with her. The bishop stayed with the Hartzells Tuesday and Wednesday until a Mr. Siler recognized him and informed Katharine. Orville came for him on Thursday and took him to the Lorenz home. Wrote the bishop, "The flood was second to Noah's."

Lorin and his family had moved from Horace Street to a house at the corner of Grafton and Grand Avenues a week before the flood; their home escaped the waters. The Wright family flood refugees stayed with Lorin three weeks and a day, when gas was restored to the Hawthorn Street house and they went back home. The removal of the sediment and the cleaning of the dirt from the cellar and lower floors and yard took a month. "Orville's automobile was submerged," wrote the bishop, "and injured several hundred dollars. Orville lost a pianola costing 500 dollars. I lost a few books of value, and the family lost two or three hundred dollars worth."

Glenn Curtiss

The Model G aeroboat was designed by Grover Loening under Orville's supervision in 1913. Katharine christened the Wilbur Wright aeroboat in 1922 in New York. After the christening Frederick H. Becker piloted poet Percy MacKaye, explorer V. Steffansson, Orville and Katharine in a flight over the Hudson.

Orville to Andrew Freedman, Chairman, Wright Company

April 11, 1913

It has now been two and one half weeks since I have been able to be in my office and we have not as yet succeeded in getting any light or heat. The water covered the first floor of my home about six feet deep, but our factory is high on the hills and far away from the water, so that there was no loss there at all. My greatest anxiety was over my own office, where I keep all our aeronautical books and papers and the scientific data upon which I base all calculations. Fire broke out in our block and destroyed the nearby buildings, but for some unexplainable reason our building, which has a shingle roof, did not catch.

My personal losses have been slight, somewhere between three and five thousand dollars. Hundreds of families and merchants in the city lost practically everything they had.

We attempted to start the factory yesterday but only succeeded in getting five of our workmen. There is some prospect that the street railway service will be resumed the early part of next week, in which case I think we will have a fairly full force.

Some damage was done to glass negatives of early flights, but the famous first flight negative suffered only slight damage. The 1903 machine was under water for several weeks in the shed of the Hawthorn Street house.

The Board of Regents of the Smithsonian Institution voted on May 1, 1913, to reopen the Smithsonian Aeronautical Laboratory and rename it The Langley Aerodynamical Laboratory. In the laboratory the wreckage of the Langley aerodrome lay under a blanket of dust.

Dr. Albert F. Zahm, professor of mechanics at Catholic University and friend of the Wrights, became the director of the Langley Laboratory. Captain Holden C. Richardson headed the advisory committee of fourteen, among whom were Glenn H. Curtiss and Orville Wright.

The 1913 Model CH was the first Wright hydroplane. Two pontoons were attached to the skids. Orville experimented with the plane on the Miami River in the summer of 1913.

A problem that plagued aeroplanes was the tendency to stall and dive toward the ground. Hap Arnold and an observer were flying over Ft. Riley, Kansas, in 1912 when suddenly the aeroplane spun around in a small circle. It then went into a nosedive. As he glanced over the aeroplane, Arnold could find nothing wrong. He pulled frantically at the controls and managed just a few feet from the ground to pull the machine out of the dive. A few minutes after he landed, hundreds of soldiers came running, expecting to find the machine wrecked and the men dead.

When Orville heard of Arnold's experience, he said that the machine had stalled and that he himself had had several like experiences. At the time of the Arnold incident Orville was working on a device to overcome the stalling feature. Lieutenant Colonel Samuel Reber, head of the aeronautical department of the Signal Corps, wrote to Orville late in 1913 about the possibility of having a manual for American aviators to show from a practical and a scientific standpoint the operation of balancing and steering a flying machine, and the dangers of certain kinds of flying. Orville was at the time at work on just such a manual, but trial testimony was preventing its completion.

Orville to Samuel Reber

Dayton, December 5, 1913

Recently no less than fifteen fatal accidents to European military aviators occurred within a space of sixteen days, yet in America they were scarcely noticed.

These accidents are the more distressing because they can be avoided. I think I am well within bounds when I say that over ninety percent of them are due to one and the same cause—*stalling. Stalling* is a term used to designate that condition of a flying machine when its speed has dropped to a point where it becomes unmanageable. Recovery is possible only by regaining speed. When in this *stalled* condition, the machine will dive downward in spite every effort of the aviator to stop it.

Stalling is a danger that can be absolutely avoided. Proper care in observing the angles of incidence indicators, such as placed by our department on its machines, will positively eliminate it, and more than ninety percent of all accidents will be done away with. Too many flyers become careless in observing the angles of incidence as shown by the indicators, but trust to their feel in gauging the angles. This they can do very well when flying on a straight course, but when the power of the motor is off and the machine is gliding downward the sense of feel often deceives them. Practically all of the accidents due to stalling occur either when a rapid climb is attempted or when the machine is gliding downward under reduced power.

Robert J. Collier, president of the Collier Publishing Company and a member of the board of the Wright Company, gave a silver trophy to be awarded by the Aero Club of America for the most significant contribution to the science of aeronautics. It was first awarded in 1911 to Glenn H. Curtiss for his invention of the hydroplane. Orville received the 1913 Collier Trophy for the development and demonstration of his automatic stabilizer. The U.S. Patent Office granted patent no. 1,075,533 for a device for maintaining automatic stability of an aircraft October 14, 1913. The patent application was filed in 1908.

Orville and nephew Horace go fishing off Kitty Hawk, 1911.

The United States Circuit Court of Appeals of New York handed down its decision in the suit of the Wrights against the Curtiss-Herring Company and Glenn H. Curtiss January 5, 1914. The court recognized the Wrights' patent as a pioneer patent and permanently enjoined Curtiss from manufacturing or selling aeroplanes in which two ailerons functioned simultaneously to produce different angles on the right and left wingtips.

Orville, wearing his usual bowler hat and business suit, and up to his knees in the beautiful Miami, checks the Model CH. He later finds the single-pontoon model gives better control of the machine.

When he received the news of the decision, Glenn H. Curtiss and his family were staying in a villa at Nice on the Mediterranean. He had gone there to demonstrate his areoboat at Venice in anticipation of an order from the Italian government.

Glenn H. Curtiss to Orville Wright

January 13, 1914

Congratulations.

Curtiss told reporters he would appeal to the United States Supreme Court, but he would pay royalties if he had to and if his appeal was dismissed, he would confine his work to areoboats and engines.

While Orville was in New York on business for the Wright Company in February, he issued a prepared statement to the press, which appeared on page one of the February 27 edition of *The New York Times*. The story said in part:

For the first time since Wilbur and Orville Wright flew in an aeroplane at Kitty Hawk ten years ago, Orville Wright told yesterday the frank and full story of what it had cost them in money, industry, and patience to launch the aeroplane as a patented and protected device.

In telling the story Mr. Wright ended a policy of silence the brothers deemed necessary at first because they feared too many men would copy their invention, and more recently because their inventions were in litigation before the United States courts.

This litigation has now proceeded, according to Mr. Wright, to the full length that the courts allow. Contrary to the contentions of Glenn H. Curtiss, Mr. Wright insists that there is no appeal to the United States Supreme Court, and therefore he considers his patents as fully established and validated.

The advice of Mr. Wright, as evolved from his own experience and that of his brother, was that any struggling young inventor should absolutely withhold all knowledge of his invention from the public, and from the patent office as well, until he has obtained $300,000 backing to be used in fighting through the tedious court processes his claims to his invention. To the law's delays and the load of worry and responsibility which these delays imposed upon him, Mr. Wright laid the illness and death of his brother Wilbur.

To his announcement that he would require Mr. Curtiss to take out a license and settle for back business, Mr. Wright added that he had personal knowledge that Mr. Curtiss had anticipated the patent decision by having all his property put beyond the reach of court seizure, and that therefore he was not hopeful that he could collect any large amount from Mr. Curtiss on infringement claims.

Mr. Wright also announced that the amount of money he would demand from aeroplane manufacturers would be in the neighborhood of 20 percent of the selling price of each machine, or $1,000 for the standard aeroplane upon the market, and perhaps $2,000 for those of especially high cost and high horsepower.

For all other aeroplane manufacturers except Mr. Curtiss, Mr. Wright said he would adopt a policy of leniency. He said that innocent purchasers of aeroplanes which were infringements would be protected and that aeroplane manufacturers who had built machines without knowing they were infringing would be dealt with lightly, but that they would either have to take out licenses or quit flying and manufacturing.

"This is not an act of harshness," said Mr. Wright, "it is an act of great benefit to aerial navigation. In the first place, all experts know, but very few laymen know, that the only advances made in aeroplane construction in the past ten years have been improvements in the motor and not to the dynamic power of the aeroplane itself.

Orville
in the 1920s

"I want all the inventors who can be possibly brought into the industry to work upon the aeroplane. And let them start with the problem of bringing in something new—beyond the scope of our patents, which will render the patents useless and obsolete. Then we will be glad to retire in their favor.

"It wasn't as if we had been fighting a stand-up foe in a square give-and-take fight. We were fighting a foe whose strategy was played in the dark. When I lay in the hospital at Fort Myer in 1908, for instance, after the fall in which I was severely injured, three gentlemen asked to look over our wrecked machine in its hanger. An appointment was made for them, but they did not appear at the appointed time, and the aeroplane was crated up. Later, when the guards were at dinner, one of the visitors (Alexander Graham Bell) took the parts of the machine out of its crate and measured each part. Later we found an aeroplane on the market duplicating the measurements of our own exactly.

"We have had to remain silent while aeroplane 'inventors' gained public applause for making aeroplanes which were exact duplicates of the machine described in our original patent papers, a type we discarded after the very first flights.

"And we had to remain silent while these men told of 'developing' this type of machine in the wilds of some unknown country. Every improvement on which we filed patent papers we found quickly added to the machines of our competitors. And, of course, our only recourse was a tedious process, beginning with an injunction and scheduled never to end until years and years later in the court of last resort. All this worried Wilbur, first into a state of chronic nervousness and then into a physical fatigue, which made him an easy prey of typhoid fever which caused his death.

"Soon after we had commenced to fly, a very fine man came to Dayton with a balloon and had engine trouble there. This was my good friend, Capt. Thomas A. Baldwin. Baldwin sent for Glenn H. Curtiss, a motor builder to come and repair his engine. We met Mr. Curtiss and told him—then an incredulous unbeliever—about our flights. We wanted him to know about them as an engine builder because we needed to have motors of the right kind. We told him all there was to tell. That was before anyone had ever flown anywhere in the world, except ourselves.

"Later, when the Aerial Experiment station was opened, with Mr. Curtiss as a demonstrator, we answered all questions asked us, even to a question as to where the centre of pressure was and centre of gravity in our machine. We were shocked when we

The Model R, the "Baby Wright," was a one-place racing machine designed for altitude and speed racing competition in 1910.

Orville poses before the 1912 Model C machine at Simms Station. One of the first five machines ordered by the Army, the military Model C climbed 200 ft. per minute, carried enough fuel for 4 hours and a weight of 450 lbs. including pilot and passenger.

found Mr. Curtiss was starting out as an exhibitor and wrote him to that effect. We received a letter saying that he had no intention of ever becoming an exhibitor or making commercial use of our invention. We offered him a license then, but he thought he could get along without it.

"We have offered to license him twice since and have agreed upon all the terms except one, which had nothing to do with the money involved. We found that one concern holding a license took off the number plates from old machines after discarding them and put the old numbers on new machines to avoid paying royalties. So we have to make our royalties apply to all repairs as well as to new machines. That is the only way to get around a question we could not otherwise settle."

Curtiss wrote to Orville March 5, 1914, suggesting he would like to meet Orville in Dayton. "A number of our mutual friends think there is an opportunity which will benefit us both financially. It seems to me that we should talk this matter over personally."

Orville did not reply to the letter. Glenn Curtiss on March 8 made a public statement. *The New York Times* story said in part:

Mr. Curtiss denies explicitly that he ever appropriated any of the Wright ideas in making his machines, and asserts his belief that his control of aeroplanes "differs fundamentally" from that used by the Wrights. He also asserts that Mr. Wright's attitude on royalties means permanent injury to aviation. Mr. Curtiss said, "In some New York daily papers there have been published certain statements attributed to Orville Wright regarding his attitude in the aeroplane patent situation. Mixed in with these direct quotations were interpolated insinuations impugning my good faith in the patent litigation, and carrying suggestions easily interpreted as such untruths as I cannot believe Mr. Wright, or any other sane man, ever made.

"The idea that any single line or part of my machine was either copied from the Wright machines, suggested by the Wrights, or by their machines, is absurd, if not malicious. My first public flights, as a member of the Aerial Experiment Association, are a matter of record, and were made months before the Wrights exhibited their machines or made their first public flights. I never had an item of information from either of the

Wrights that helped me in designing or constructing my machines or that I ever consciously used. I believe today, as I have always believed, that the Curtiss control differs fundamentally from that employed by the Wrights and that its superiority to the Wright system is demonstrated by the records of two machines during the past five years. That I was unable to satisfactorily demonstrate this intricate technical point to the court I consider a misfortune largely due to the fact that our knowledge of aviation was vastly less when this case went into court several years ago than it is today.''

Mr. Curtiss concluded by quoting editorial comment from *The Boston Transcript:* ''As a matter of fact, Mr. Wright very evidently has a firm intention of bottling up the aviation industry in this country, and confining the manufacture of aeroplanes solely to the Wright plant in Dayton. It is difficult to believe that Mr. Wright is actuated by other than personal animosity. The effect of the Wright decree will beyond question numb what little life remains today in aviation in America. One of the most progressive and wide-awake concerns in the country, the Curtiss Aeroplane Company of Hammondsport, N. Y., is virtually forced out of the country, a situation with which Mr. Wright is highly content, as appears from the text of his interview. Only the production of a non-infringing type, as Mr. Wright suggests, can hope to stimulate in this country the interest in aviation Mr. Wright apparently is so determined to kill.''

The 1913 Model E was a one-place exhibition aircraft, basically a small Model C but with a single pusher propeller. That's Orville flying the Model E.

The Aero Club of America extended in 1906 the first official congratulatory resolutions in the United States to the Wright brothers for their achievements in powered flight. President Taft in 1910 presented the Aero Club gold medals to the brothers.

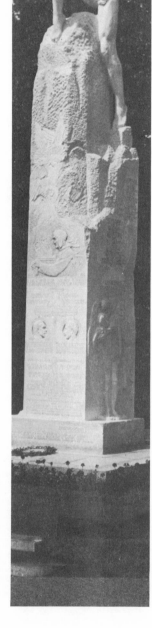

The Wilbur Wright Memorial was dedicated July 17, 1920, in Le Mans. Commodore Louis D. Beaumont of Dayton commissioned sculptor Paul Landowski and architect Paul Bigot to design and build the 40-foot monument. At the laying of the first stone of the monument in 1918, Beaumont presented a bronze wreath on behalf of the city of Dayton. Baron Paul d'Estournelles de Constant in 1912 began the subscription campaign for funds to build the monument to commemorate Wilbur's flights in France in 1908.

The Coupe Michelin pour L'Aviation, designed by Paul Roussel, commissioned by Edward and Andre Michelin, was awarded to Wilbur in 1908 for his flight of 123 kilometers, 200 meters in 2 hours, 18 minutes, 33-3/5 seconds. The Cup is on display at Wright State University, Dayton.

13

Silent Truth

I had thought that truth eventually must prevail, but I have found that silent truth cannot withstand error aided by continual propaganda.
—Orville Wright, 1928

Dr. Charles D. Walcott, who succeeded Samuel P. Langley as secretary of the Smithsonian Institution, said in a letter to the Curtiss Aeroplane Company early in 1914 that it seemed desirable to make a thorough test of the principles involved in the construction of the Langley 1903 machine. Walcott asked Glenn Curtiss if he would be willing to estimate the cost of restoring the Langley aerodrome and make the tests. Curtiss replied in a letter on March 16, "I believe that this would be a good thing to do, and I think I can get permission to rebuild the machine, which would a long way towards showing that the Wrights did not invent the flying machine as a whole but only a balancing device, and we will get a better decision next time."

Secretary Walcott called Alexander Graham Bell March 26 and told him that Curtiss was anxious to make a copy of the Langley aerodrome and that he, Curtiss, would have it ready to try in May. Walcott said that Curtiss thought it would cost no more than $2,000 and asked if Bell would approve the appropriation from Smithsonian funds. Bell told Walcott that he thought it would not be advisable to use Smithsonian money. Walcott then said he would chip in $1,000 and Bell said he would give money. "It would be much better." said Bell, "to have the experiment made by outside parties at their own expense than have the Smithsonian responsible for it."

Walcott felt that Curtiss should make the experiment at Hammondsport without directing any attention to it, and that if it were successful he could repeat it publicly on Langley Day.

Walcott decided to repair the original Langley aerodrome, which was shipped in a boxcar to Hammondsport in April 1914. The Hammondsport newspaper reported the arrival of the Langley machine.

The Smithsonian assigned Dr. A. F. Zahm to represent it at Hammondsport, and Walcott hired Charles Manly, Langley's pilot when the aerodrome sank into the Potomac in 1903, to repair the engine. Plans were first to show whether the original aeroplane was capable of a sustained free flight with a pilot and second, to determine more fully the advantages or disadvantages of a tandem-wing type of aeroplane. If the restored machine flew, plans were to install a 1915 motor.

Curtiss flew the restored Langley machine on May 28, 1914, taking off from the surface of Lake Keuka. "Many eager witnesses and cameramen were at hand, on shore and in boats," Zahm reported. "The four-winged craft, pointed somewhat across the wind, went skimming over the wavelet, then automatically headed into the wind, rose in level poise, soared gracefully for 150 feet and landed softly on the water near the shore. Mr. Curtiss asserted that he could have flown further, but being unused to the machine, imagined the left wing had more resistance than the right."

Alexander Graham Bell sent a telegram to Curtiss congratulating him on his successful vindication of Langley's aerodrome. Walcott ordered a Curtiss 80-hp engine installed in the aeroplane for the second series of experiments, to prove the practicability of tandem wings.

A reporter from *The New York Times* said in his story that "aside from the floats the machine was almost exactly as it was when the faulty launchings were made over the Potomac River." The reporter was unaware that Curtiss had made thirty significant changes, "using knowledge of aerodynamics discovered by the Wrights," said Fred Kelly in *The Wright Brothers,* "but never possessed by Langley."

Griffith Brewer, friend of the Wrights and their legal representative in England, gathering material for a history of aviation, visited Orville at the Wrights' new home, Hawthorn Hill, in Dayton in May 1914. He told Orville that he had read in the American press about the Curtiss trials at Hammondsport, which were presumably being made in

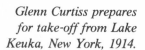

Glenn Curtiss prepares for take-off from Lake Keuka, New York, 1914.

order to prove that the Wrights were not pioneers in flying. Orville suggested that Brewer should go to Hammondsport to see what was going on.

At Hammondsport Brewer posed as a newspaper correspondent from England and watched the changes being made for the second-phase experiments.

"Glenn Curtiss and Zahm were fitting the Langley machine with floats," said Brewer in *Fifty Years of Flying,* "and just as one alteration leads to another, it was found necessary not to make just one or two changes, but in the end to make over thirty changes in the structure, besides introducing alterations which the aeronautical knowledge of 1914 dictated as necessary beyond the Smithsonian knowledge of 1903. For instance, by this time they knew that Langley had mistaken the position of the centre of pressure and that alone would cause the wings to collapse, so the position of the staying to support the wings was moved back about three feet. It was known also that the deep camber of the Langley machine was less efficient than the small camber of 1914, consequently the wings were changed from a camber of one in twelve to a camber of one in eighteen. This change reduced the area of the wings by fifty-two square feet."

On his return to Dayton Brewer reported to Orville what he had seen and also that he had written to *The New York Times* a letter (published June 22, 1914), in which he listed numerous changes which had been made before the Langley aeroplane was flown at Hammondsport. His letter concluded, "Why, if such a demonstration was considered desirable, why was not the old historic relic left untouched and a copy made to satisfy an insane curiosity? Why not an impartial, unprejudiced person chosen to make the test, instead of the person who has been found guilty of infringement of the Wright patent?"

Orville's brother Lorin went on an observation mission to Hammondsport. He used the name W. L. Oren.

Lorin Wright's Notes

I walked around toward the camp near the shore and came to a stream or lagoon which I could not cross. I watched the proceedings from this point through a pair of glasses which I had taken along with me from Dayton. The distance from the launching place was, I think, about 600 feet.

About ten o'clock Mr. Walter Johnson mounted the machine and started the motor. The machine gradually acquired speed and after running as near as I could judge about 1,000 ft. the rear wings broke about midway the length of the spars and folded upward. They also broke or pulled from sockets at points where they joined to frame. When machine stopped the outer edges dropped and dragged in the water.

I immediately walked up to bridge and over to grounds. When I arrived at the hangars the machine had just been towed to shore and the workmen were trying to bring it up to the runway. I took four pictures of the machine when some of the workmen noted the fact and notified Mr. Johnson. He demanded that I should give up the film. At first I refused and started to leave the grounds. Mr. Johnson and several others left the machine and came running up to me demanding the film saying that they could not allow any pictures of the wrecked machine to be made. I replied that the day before they had told me I could take all the pictures I wanted.

At this juncture a man who I am informed was Mr. Henry Weyman came up and apparently took charge of the situation. He insisted that I should not leave the grounds until I had delivered up my pack of films. I asked him why. He replied that on account of "legal complications" they wanted no pictures of the machine in its present condition to get out. I finally yielded the film pack and he insisted on supplying me with another to replace it.

When the 1914 annual report of the Smithsonian Institution was published August 1915, it not only contained Albert Zahm's description of the flight of May 28, but also a statement that the aerodrome of Samuel P. Langley "has demonstrated that with its original structure and power, it is capable of flying with a pilot and several hundred pounds of useful load. It is the first aeroplane in the history of the world of which this can be truthfully said."

The statement led to a controversy between Orville Wright and the Smithsonian Institution that continued for more than quarter of a century.

Orville continued to experience trouble with his leg and hip injured in the 1908 crash in which Selfridge had been killed. Because of bouts of pain and the necessity of periods of bed rest and because he was more interested in research than business, Orville sold his shares in the Wright Company in October 1915 to a syndicate. He was retained as consulting engineer at an annual salary of $25,000.

While he was driving in early December 1915, Orville, who was forty-four, felt severe back pain and was taken to Hawthorn Hill, where he spent nearly eight weeks in bed. When he recovered, he moved his office and workshop to 15 North Broadway Street. One of his projects was to renovate the 1903 machine for an exhibit at the Massachusetts Institute of Technology June 11 through June 13, 1916, where he and Alexander Graham Bell were honored for their achievements in applied science.

That summer Orville, Katharine, and Bishop Wright spent July 11 to September 3 vacationing at Georgian Bay, Ontario, Canada. In September he bought Lambert Island, where he afterwards entertained his family and friends for many years. The island of about twenty-six acres was mostly rock. On its highest point it had a main cottage with a living room, kitchen, one bedroom, and screened porch. Guests at the island were provided with cottages, and meals were served in the main house. Orville was the cook. He permitted no one to enter the kitchen while he was cooking. He made toast between two sheets of metal held over an oil stove. He was good at soups, vegetables, lamb chops, coffee, and dessert.

Although Georgian Bay is a prime fishing area, Orville fished little, but he kept the gasoline motors in order and built new cribs for the boat slip. He rigged up a gasoline motor to pump water from the lake to a water tower from which water was brought by gravity to the kitchen. He built a second-story porch on the main cottage and installed a bathroom. When he built a tool house, he invited his guests, including Griffith Brewer, to help him, providing hammers and more than enough nails.

Bishop Wright's Diary

Georgian Bay, July 12, 1916

Went on a sail with Orville, three miles on the Bay. I slept a nap; dined before 1:00. Slept an hour in the afternoon. A skipper rowboat brought Miss Williams who brought us some milk.

July 17, 1916

Orville and Katharine went on the supply boat. I am 4 months less than 88 years old. The Williams girls and Mrs. Hodgins came and brought their knitting. People knit for the hospitals up here.

July 21, 1916

Rev. Reinhold gave us some fish. Nobody came. Horace (Lorin's son) caught a pike 29½ inches long.

July 31, 1916

Katharine looks after a supply boat; finds it at last. Orville and the boys go hunting after minnows. We find them better than fishworms to catch black bass.

August 16, 1916

We went out boat-riding and our motor irresistibly stuck, and Orville paddled home a mile and a half.

August 19, 1916

This is Orville's and Katharine's birthday—he 45, she is 42 years old. She resembles her grandfather Koerner; he more like the Reeders.

Dayton, September 15, 1916

Orville got me a bright light for my room in the evening, to read by. My central light is too high and the others too dim and to one side.

September 17, 1916

I lack two months of being 88 years old. Mr. Edward A. Deeds came while I was asleep in the afternoon, and Orville and Katharine went away with him. I have been inclined to sleep today.

October 22, 1916

Lorin's family, including Ivonette, dined with us. Reuchlin's also. They staid through the afternoon. Ivonette sung and Lulu played for her. We had many "Victor" songs. I ate supper and went to my room. I took up the Third Chapter of Romans (1-80) which I have not been able to understand.

January 1, 1917

This is my 89th New Year. Lorin and Netta, Ivonette and Leontine spent evening with us.

From Monday, January 8, to Saturday, March 10, 1917 there were fewer diary entries of activities than days with no entries at all. Then

Saturday, March 10, 1917

Scipio came. He weighs 16 pounds. He is a St. Bernard dog. He is a good-looking puppy.

When the bishop did not appear for breakfast on the morning of April 3, Carrie Grumbach, the housekeeper, told Orville that she was worried about the old gentleman. Orville hurried to the bishop's room and found that his father had died in his sleep. Services were held at Hawthorn Hill on April 5 and he was buried in the family plot in Woodland Cemetery.

Five days after Congress declared war on Germany April 6, 1917, the Dayton-Wright Airplane Company was organized by Edward A. Deeds, Charles F. Kettering,

Bishop Milton Wright
1828—1917

The Liberty engine for De Havilland 4 was built in the Dayton Wright Airplane Company factory, 1917.

Harold E. Talbott, Sr., and H. E. Talbott, Jr. Orville Wright was consultant and engineer for both the Dayton Wright Airplane Company and Wright Field Company, which was incorporated at the same time. The United States government bought the Delco-Light plant and turned it over to the Dayton-Wright Airplane Company for the manufacture of Liberty engines to power De Havilland 4's. "Through the influences of these men," Orville said to Glenn L. Martin, "the government has leased 2,500 acres surrounding the old field near Dayton, and is going to establish a training camp of four squadrons. This, I believe, will be the largest camp in America."

Orville wrote to Griffith Brewer, "I am just now getting my wind tunnel in working order. I do not know what I will have to do in the military work. I have been commissioned a major in the Aviation Section, Signal Officers Reserve Corps. It is possible I may have to go to Washington, although General Squier told me a short time ago they thought they would have more need of me in the laboratory."

In June of 1917 when American troops began combat training in France, Orville wrote, "When my brother and I built and flew the first man-carrying machine, we thought that we were introducing into the world an invention that would make further wars practically impossible. Nevertheless, the world finds itself in the greatest war in history. Neither side has been able to win on account of the part the aeroplane has played. Both sides know exactly what the other is doing. The two sides are apparently nearly equal in aerial equipment, and it seems to me that unless present conditions can be changed, the war will continue for years.

"However, if the Allies are equipped with such a number of aeroplanes as to keep the number of enemy planes entirely back of the line so that they are unable to direct gunfire or to observe the movement of Allied troops—in other words, if the enemy's eyes can be put out—it will be possible to end the war. But to end the war quickly and cheaply, the supremacy in the air must be so complete as to entirely blind the enemy."

Orville to W.R. Jackson

Dayton, July 19, 1917

I note what you say in regard to the use of cayenne pepper in bombs. Yes, I have myself had some experience with red pepper. When I was a small boy, several of my schoolmates and myself conspired to get out of school by dropping a package of red pepper down the register. Unfortunately the pepper did not get hot enough, but a day or two later, when it did begin to burn, the teacher simply said that she was sorry, opened the windows, and we sat there and sneezed and wiped our eyes.

However, I do not think that this forms a gas as terrible as some of those that are now used on the battlefields. But I do not think in any case this is the way the war is to be won. Since the first use of the gasses in which the Allies were unprepared and taken by surprise, they have had but little effect.

Orville to Frank Harris

Dayton, August 1, 1917

An attempt to destroy the Krupp works at Essen could be undertaken successfully only in case the Allies have a great preponderance of fighting aeroplanes, so that the machine carrying bombs could be safely convoyed. I have never been a strong advocate of bombing from aeroplanes. I certainly would not like to see the Allies adopt the Germans' barbarous policy of dropping bombs among the civilians where no military advantage is to be gained by it.

The important task that lies before our country is that of supplying the Allies with good aeroplane motors. This is a very difficult job. If it were not so difficult the Allies would by this time have been equipped to supply themselves. I do not think that we will be able to begin production on a large scale much inside of a year, but at the end of that time, with the manufacturing facilities which we have in America, we should be turning out motors at a very rapid rate.

I believe the total number of Allied machines on the front is not more than three thousand or three thousand five hundred. I have never understood that they had nine machines in reserve for every machine on the front. If ten thousand more machines could be put on the front, each provided with a well-trained flyer, you can see that the Allies would have a tremendous advantage, and should be able to drive the German machines from the skies.

Just before the Armistice Griffith returned to Dayton from England after a four-year absence caused by the War. He found that the campaign of belittling the Wrights was still on. Many Americans who had set out to expose the destructive propaganda used by the Smithsonian sooner or later dropped the investigation because they realized that the Smithsonian was the leading scientific authority and that anyone attacking that body would do so with the disapproval of orthodox scientists in America.

Brewer found that the Langley aerodrome had been placed on exhibition in the Smithsonian in 1918. The label read:

ORIGINAL LANGLEY FLYING MACHINE
1903

The first man-carrying aeroplane in the
history of the world capable of sustained
free flight. Invented, built, and tested
over the Potomac River by Samuel Pierpont
Langley in 1903. Successfully flown at
Hammondsport, N. Y., June 2, 1914.

*Orville and Griffith
Brewer at the Smithsonian*

"On my summer visit to Orville in 1921," Brewer later wrote, "it occurred to me that I could afford to attack the Smithsonian when prominent Americans could not, and I suggested that I should extend my visit on that occasion and prepare a paper which I could read before the Royal Aeronautical Society on my return to England. To this Orville agreed and we settled down to a methodical analysis of the differences between the original Langley machine and the machine experimented with at Hammondsport.

"I thoroughly enjoyed the two months' investigation and the work at the drawing board in Orville's laboratory. With his invaluable help and the help of Miss Mabel Beck in typing the paper I sent a copy to the Council of the Royal Aeronautical Society.

"Lord Montagu of Beaulieu presided when I read my paper on the 20th October 1921, but before reading the paper copies of it were sent to Walcott, Zahm, Manley and Glenn Curtiss. The paper created a tremendous commotion in the aeronautical world. Leading articles appeared in the London *Times* and the *New York Times* and also in the London *Daily Mail* and *Daily Telegraph*.

"Everyone thought that the lecture would be followed by a libel action, and C. G. Grey hastened to discuss the issues, fearing that the subject would be closed for discussion immediately a libel action had been started. But I was sure of my ground. No doubt there was some risk of an action being brought against me, but I should have had no doubt as to the issue and was sure I would be able to prove I had in the public interest exposed a fraud.

"One would have thought that this exposure would silence propaganda supporting the false contention that Langley's original design had been flown at Hammondsport, while concealing the fact that a large number of alterations had to be made in order to enable it to rise from the water; yet this propaganda continued on the principle that if you make a statement long enough the public will ultimately believe it."

On the twentieth anniversary of the 1903 flight, Orville was asked to prepare a speech dealing with twenty years of progress in aviation and to give it over radio station WLW in Cincinnati. Although Orville wrote the address, he refused to go to Cincinnati to read it. Dr. D. Frank Garland of the National Cash Register Company agreed to read it for him.

When it was time for the speech, Orville walked down his hill to the Hall home. The Halls had a radio; he did not. The radio had two headphones. Mr. Hall put one on and gave the other to Orville. A few minutes after the speech had begun, Mr. Hall took off his earphones and stared at Orville. "How in the world can I be here listening to you in Cincinnati when you are right here in this room listening to yourself?"

Orville listened to himself recounting the history of the four flights and also saying, "Twenty years ago my brother and I thought that its use would be principally scouting in warfare, carrying mail and other light loads to places inaccessible by rail or water, and sport. But the wildest stretch of the imagination of that time would not have permitted us to believe that within a space of fifteen years actually thousands of these machines would be in the air engaged in deadly combat. Our expectation of its value in scouting has been fulfilled. Surprise attacks which formerly won battles are now impossible. But we did not foresee the extent to which the airplane might be used in carrying the battle line to the industrial centers and into the midst of non-combatants, though we did think it might be used in dropping an occasional bomb about the heads of the rulers who declared war and stayed at home. The possibilities of the airplane for destruction by bomb and poison gas have been so increased since the last war that the mind is staggered in attempting to picture the horrors of the next one. The airplane, in forcing upon governments a realization of the possibilities for destruction, has actually become a powerful instrument for peace."

Orville had for two years been mulling over the suggestion that Griffith Brewer had made to him. In late 1923 Orville wrote to Brewer, "If you are right in thinking that the officers of the Museum would be keenly interested in securing this (1903) machine for exhibition in the Kensington Museum, I should be inclined to let them have it. The fire risk are too great where it is now stored and I have no better place for it.

"Of course the machine ought to be in the National Museum at Washington."

On April 30, 1925, Orville stated to the *Dayton Daily News* and the Dayton *Journal* that he intended to give the original Wright 1903 aeroplane to the Kensington Science Museum in London.

Orville's public statement stung Smithsonian Secretary Walcott, who defended the Institution in a special statement in *Outlook,* a weekly magazine. He reaffirmed the correctness of the 1918 Smithsonian label on the Langley machine and after reviewing the trials, he said, "These tests were no more initiated for the purpose of influencing patent litigation than were Langley's original experiments."

In an accompanying editorial, *Outlook* said, "The Langley aeroplane did not have an impartial trial. That trial was conducted by men hopeful of winning for Professor Langley an even greater fame than he achieved in his lifetime, and by men directly interested in proving that Orville and Wilbur Wright were not the first designers of a practicable man-carrying plane."

On May 29, 1925, Orville announced that he was agreeable to keeping the original Wright machine in the United States if the Smithsonian Institution labelled truthfully the Langley machine, if the Institution published both sides of the Langley controversy in its annual report, and if the Wright machine was labelled the first man-carrying machine in the world.

Within a few days Secretary Walcott asked Joseph S. Ames, professor of physics at John Hopkins, and Rear Admiral David W. Taylor to study the Langley machine and its history and advise him whether they thought the label should be changed. Ames and Taylor suggested that the label be changed to:

LANGLEY FLYING MACHINE
ORIGINAL LANGLEY FLYING MACHINE
OF 1903, RESTORED

In the opinion of many competent to judge, this machine was the first heavier-than-aircraft in the history of the world capable of sustained free flight under its own power, carrying a man.

This machine slightly antedated the Wright machine designed and built by Wilbur and Orville Wright, which on December 17, 1903, was the first in the history of the world to make a sustained free flight under its own power, carrying a man.

Walcott changed the label but he did not change Orville's mind or intention.

While the Smithsonian controversy continued, the United States War Department honored the Wright brothers by naming the new Army Air Service Field at Dayton "Wright Field." This broke an Army precedent that no military flying field was to be named in honor of any living man. Orville and Katharine attended that ground-breaking ceremonies at Wright Field, April 13, 1926. Katharine had not resumed teaching after Orville's crash at Ft. Myer but had travelled extensively with Orville, entertained guests at Hawthorn Hill, managed the household, participated in Dayton civic activities and in the Oberlin College Alumni Association. It was at Oberlin that she renewed a friendship with a classmate Henry J. Haskell.

One day in 1926, Katharine, who was fifty-two, told Orville that she was going to marry Haskell. The news stunned Orville. He had never thought of the possibility of Katharine's leaving him or Hawthorn Hill. If she married, she would move to Kansas City because Haskell was the associate editor of *The Kansas City Star*.

She was married November 20, 1926, at Oberlin by the president of the college, Henry Churchill King. Orville did not attend the wedding.

He never spoke to her again or saw her until she was dying.

Katharine and Orville on shipboard coming home from Rome, 1909.

Katharine Wright Haskell
1874—1929

On May 20, 1927, Orville went to the home of his niece and nephew, Ivonette and Harold Miller. Orville still did not own a radio. He sat down by the Millers' radio, propped his elbows on his knees and remained all day without stirring. Noreen, the maid, asked if he would like a glass of water or a cold drink, but he shook his head without speaking. Charles A. Lindbergh was flying across the Atlantic to Paris.

Orville to Major Albert B. Lambert

Dayton, June 9, 1927

The people of Dayton are most anxious to have Colonel Lindbergh come here as the guest of the city immediately after the celebration at St. Louis. I believe Dayton has had a longer and more intense interest in aviation than any other city of our country. For this reason it is able to appreciate to the full Colonel Lindbergh's wonderful flight. I am sure no city will give him a more cordial welcome.

I am writing to you as an old friend in this art, hoping you will be glad to use your influence in giving Dayton her wish.

Personally I have more than ordinary interest in Colonel Lindbergh; first, because he has so strikingly demonstrated the possibility of an art which I had a part in founding, and second because his conduct in the midst of overwhelming popularity has been such as to command the admiration and respect of everyone.

Please extend to Colonel Lindbergh my most cordial greetings with the hope that I may soon have the pleasure of his personal acquaintance.

*Orville and Major Curry
greet Colonel Lindbergh
at Wright Field
June 18, 1927.*

Lindbergh arrived in Washington June 11. President Coolidge greeted him at the Washington Monument. New York staged a demonstration for him on June 13, 14. He then went, as he had promised, to St. Louis for a celebration and then flew to Wright Field on June 18 to pay his respects to Orville Wright. Orville and Major John F. Curry met him at the field.

A crowd had gathered in downtown Dayton, expecting to see the new world hero. But Lindbergh said he would not be driven through the crowds. The car in which he rode wound through the back streets to Hawthorn Hill, where there was to be a small dinner and peace and quiet.

The crowds, disappointed in being unable to see Lindbergh, arrived at Hawthorn Hill, swarming over the grounds, trampling shrubbery, climbing trees, surging up on the portico, all shouting for Lindbergh.

Inside the house Orville and Lindbergh and the guests fled upstairs. When Orville saw what was happening to the lawn and the house, he asked Lindbergh if he would be willing to appear on the second-floor balcony. Lindbergh made a very brief appearance with Orville at his side, and the satisfied crowd left. Lindbergh slept that night in the room that had been built for Wilbur.

The grass and shrubbery at Hawthorn Hill had barely recovered from Lindbergh's admirers when long-time friend Griffith Brewer arrived. He and Orville sat in the library chatting after dinner. "I have often thought," said Brewer, "that after you and your brother learned to fly, the problem that had baffled men for centuries suddenly seemed most simple. You would think anyone could have done it. There is a passage of poetry that expresses that idea. I have been trying to think of it for years. All I can remember is 'so easy it seemed once found, which yet unfound most would thought impossible.' There is more to it about invention. I wish I could find the whole passage. Do you know it?"

"No, I don't," answered Orville, "but I have an extensive collection of poetry here in the library. Let's look." The two men spent an hour hunting for the lines, but the poetry eluded them.

The very next morning one of the coincidences so common in life occurred. In a letter asking for Orville's autograph the writer included the very quotation Brewer had asked about and gave the information that it came from *Paradise Lost,* Book VI. Orville took down his volume of Milton and began to search. Finally at line 499 he came to the passage, which began

> *Th' invention all admired, and each how he*
> *To be th' inventor missed;*

It concluded as Brewer had remembered. Orville put the book back on the shelf, pulling the book directly above it out from the shelf a quarter of an inch.

After dinner that night Orville said, "I'd like to find the passage of poetry we talked about last night. I have never mentioned this before, but I am somewhat psychic. One night years ago when I had gone to bed, I heard Wilbur coming up the stairs. Something told me that instead of turning off the gas jet, he had merely blown out the light. I went downstairs to check. Sure enough, the light was out, but the gas jet was hissing. We could have been blown up."

"Remarkable."

"I thought I might try to locate the passage by using my psychic powers. I'll blindfold myself, run my fingers along the books and perhaps my psychic genius will guide them to the book."

After blindfolding himself, Orville ran his fingers along the shelves. His fingers stopped at one book and he pulled out the volume. He took off the blindfold. "H'mmmm. Milton. Something tells me this is the book."

Brewer looked at the book. "Milton? It doesn't sound like Milton to me."

"There must be a reason why my fingers were led to this book." Orville leafed through the pages long enough to make his act look good. Then he handed the volume to Brewer and pointed to the lines. It was a long time before Orville confessed.

Before Brewer left Hawthorn Hill, he renewed his suggestion to Orville to send the 1903 machine for display in the Kensington Museum. Orville made up his mind. He sat down at his desk and wrote for later publication the reasons for what he was about to do:

While Professor Langley was secretary of the Smithsonian all of the relations between that Institution and ourselves were friendly. At that time Wilbur and I were universally given credit not only for having made the first flight, but for having produced the first machine capable of flight, and for the scientific research from which this first machine sprang.

After Professor Langley's death the attitude of the Smithsonian began to change. The Institution began a subtle campaign to take from us much of the credit then universally accorded to us and to bring this credit to its former secretary, Professor Langley. Through some clever and some absolutely false statements it succeeded in doing this with people who were not acquainted with the facts.

To illustrate the kind of thing to which I object in the attitude of the Smithsonian, I will cite out of many a few specific cases:

It misrepresented in the *Annual Report* of the secretary for the year 1910 (page 23) the statement made by my brother Wilbur at the time of the presentation of the Langley Medal to us by inserting a quotation not used by him on that occasion, but used in a different connection at another time. The improper use of this quotation created a false impression over the world that we had acknowledged indebtedness to Langley's scientific work; that it was Langley's scientific work and our mechanical ingenuity that produced the first flying machine. This was not true.

Our original 1903 machine was offered in 1910 to the Smithsonian for exhibition in the National Museum. The officials did not want it, but preferred a much later model of less historic interest.

After the United States Circuit Court of Appeals had given a decision pronouncing Glenn H. Curtiss an infringer of the Wright invention and recognizing the Wrights as "pioneers" in the practical art of flying heavier-than-air machines, Curtiss was permitted to take the original 1903 Langley machine from the Smithsonian to make tests in an attempt to invalidate this title of "pioneer," for the purposes of another law suit. The Smithsonian appointed as its official representative at these tests the man who had been Curtiss' technical expert in the former suits and who was to serve again in that capacity in a new one. It paid Curtiss $2,000 towards the expense of the tests.

It published false and misleading reports of Curtiss' tests of the machine at Hammondsport, leaving people to believe that the original Langley machine, which had failed to fly in 1903, had been flown successfully at Hammondsport in 1914, without material change. These reports were published in spite of the fact that many changes, several of them of fundamental importance, had been made at Hammondsport; among which were the following: Wings of different camber, different area, different aspect; trussing of a different type, placed in a different location; Langley's fixed keel omitted; motor changed by substituting different carburetor, different manifold, and different ignition; propeller blades altered; hydroplane floats added; wing spars, which collapsed in 1903, reinforced; tail rudder made operable about a vertical axis, and connected to a regular Curtiss steering post; small vane rudder replaced by a large rudder of different design.

This machine restored back to its original form with much new material, the old having been mutilated or destroyed at Hammondsport, was placed in the National Museum with a false label, saying that it was the first man-carrying aeroplane in the history of the world capable of sustained free flight, and that it had been successfully flown at Hammondsport, June 2, 1914.

Following the controversy on the subject three years ago (1925) the old label was removed and a new one still containing false and misleading statements was put in its stead.

In spite of this long-continued campaign of detraction, for years I kept silent, with the thought that anyone investigating would find the facts and expose them. I had thought that truth eventually must prevail, but I have found silent truth cannot withstand error aided by continued propaganda. I have endeavored to have these matters investigated within the Smithsonian itself. I wrote to the Chancellor of the Institution (Chief Justice William Howard Taft) asking for an investigation of the acts of its Secretary in this matter, and received an answer that while the Chancellor nominally was the head of the Board of the Smithsonian Institution, his other duties were such as to make it impossible for him to give any real attention to the questions which have to be settled by the Secretary. I have publicly expressed the wish that some national scientific society or other disinterested body make an impartial investigation of my charges against the Smithsonian. To this there has been no response.

In sending our original 1903 machine to the Science Museum, London, I do so with the belief that it will be impartially judged and will receive whatever credit it is entitled to. I regret more than anyone else that this course was necessary.

Orville, on January 31, 1928, shipped the 1903 machine to the Kensington Museum in London.

British schoolboys admire the Wright 1903 Flyer at the Kensington Museum, London.

14

Reveille

Wake: the silver dusk returning
Up the beach of darkness brims,
And the ship of sunrise burning
Strands upon the eastern rims.
                         —A. E. Housman, *Reveille*

On a bright spring day, March 28, 1928, King George V and Queen Mary drove to the Kensington Science Museum to open the new galleries. After the ceremonies the King unveiled the Wright 1903 flying machine. When newspapers reported that thousands queued up to see the aeroplane, questions were asked in the press, both in England and in America, why the original machine—the first to fly in the history of the world, although invented and flown in America—should be in the charge of the Science Museum in London.

Public opinion in America prompted Dr. Charles Greeley Abbot, who became Secretary of the Smithsonian in January 1928 on the death of Secretary Walcott, to invite Orville Wright to a personal conference on April 19. At their luncheon meeting Orville told Abbot his side of the controversy and stated that he thought that the Smithsonian would have to correct history.

After his discussion with Orville, Abbot wrote "The Relation Between the Smithsonian Institution and the Wright Brothers." In it he said that everybody agrees that the Wright brothers made the first sustained human flights in a power-propelled heavier-than-air machine. He said that Alexander Graham Bell in his 1910 "Historical Address" had made less prominent in comparison with Langley's achievements the successful pioneer work of the Wrights "than he might well have done appropriately on that occasion."

He accepted Orville Wright's assurance that the comments by Wilbur about Langley had not been made at the Langley presentation and conceded that putting the tests of the Langley plane into the hands of Curtiss showed lack of consideration.

The 1903 Wright Flyer on display in the South Kensington Science Museum, London

Dr. Abbot expressed regrets on behalf of the Smithsonian Institution:

1. That any loose or inaccurate statements should have been promulgated by it which might be interpreted to Mr. Wright's disadvantage.

2. That it should have contributed by the quotation on page 23 of the Smithsonian *Annual Report of 1910* to the impression that the success of the Wright brothers was due to anything but their own research, genius, sacrifice, and perseverance.

3. That the experiments of 1914 should have been conducted and described in a way to give offense to Mr. Orville Wright and his friends.

Abbot renewed to Orville the earlier invitation of the Smithsonian of March 4, 1928, to place for perpetual preservation in the United States National Museum the 1903 machine. He also directed that the labels on the Langley aerodrome should be modified so as to tell nothing but facts, without additions of opinion as to the accomplishments of Langley.

After Orville read Abbot's monograph, he stated for publication that Dr. Abbot's publication was "mostly a hollow gesture and scarcely made a start towards clearing up the serious matters in the controversy."

"In twenty-five years flight has been extended from one minute to more than sixty-five hours; from one-half mile to more than four thousand miles and from a few feet above the ground to more than one-half mile above it," said Orville at Kitty Hawk in a statement on the twenty-fifth anniversary of the 1903 flight. "The great strides made in aviation in the last two years would indicate that we have not yet even approached the limit of possibilities."

The honors paid to the Wrights on the twenty-fifth anniversary could not be crowded into one day. On December 4, 1928, Representative Lindsay C. Warren of North Carolina, introduced House Resolution 332 to appoint a Congressional committee to attend the ceremonies honoring the twenty-fifth anniversary. December 10 Orville was the guest of honor of the city of Dayton and delegates to the International

Civil Aeronautics Conference to be held in Washington December 12 to 14, came to Dayton to pay respects to Orville and to be present at the laying of a wreath on the grave of Wilbur. At the opening session of the conference Orville, a delegate to the conference heard himself praised by President Calvin Coolidge. The same evening German Ambassador and Madam Friedrich W. von Prittwits entertained Orville at the Embassy. On the same day the United States Post Office issued two-cent and five-cent postage stamps in commemoration of the first flight. The two-cent stamp pictured the first Wright machine in flight.

The Department of Commerce showed a special motion picture history of aviation on December 13 and on the same day Orville escorted Colonel Lindbergh to the platform at the Chamber of Commerce Building to receive the Harmon Trophy. The following day Orville was guest of honor at the Mayflower Hotel dinner for the entire conference. A public reception given by the United States Senate honored Orville the next day. He was introduced to the Senate by Vice President Charles G. Dawes.

December 16 the steamer *District of Columbia* took Orville and two hundred distinguished guests from Washington to Norfolk en route to Kitty Hawk, his first trip there since 1911.

The Kitty Hawk ceremonies included the laying of the cornerstone of the national memorial to the Wright brothers on Kill Devil Hill and the unveiling of the granite monument, which had been placed by the National Aeronautic Association on the sands from which the brothers had made their December 1903 flights.

At the same time in London the Royal Aeronautic Society held a dinner to honor the Wrights. The tables in the Science Museum at Kensington were arranged around the Wright 1903 machine.

Congress the next day voted to present to Orville and posthumously to Wilbur the Distinguished Flying Cross. In presenting the award to Orville, Secretary of War Dwight F. Davis said, "Mr. Orville Wright, by his vision, perseverance, courage and skill, in

The day that Orville took this picture of E.C. Huffaker, Octave Chanute, George Spratt, and Wilbur in the camp in 1901, they had no idea that near the site the Wright Brothers National Memorial would one day stand. A replica of the shed is displayed at Kitty Hawk.

collaboration with his brother Wilbur Wright designed, constructed, and operated the airplane first at Kitty Hawk, North Carolina, on December 17, 1903, made the first successful flight under its own power and carrying a human operator, thereby making possible the achievements which are now stirring the emotions and pride of the world.''

Orville said, "Thank you."

So many honors followed the twenty-fifth annniversary that possibly all that Orville could say was "Thank you."

"Orville and Wilbur were so original and bold," Emil Ludwig wrote in 1928, in naming Orville as one of the four greatest living Americans, "that they might have come from the pages of Homer. The sublime quality in Wright is, after all, not the lightning flash of genius; it is the immensity of perseverance, the sure faith in reaching the sought-for goal, and the courage to rise again and again."

The Indiana State Legislature established the Wilbur Wright Memorial Commission to accept and maintain the Wilbur Wright birthplace near Newcastle. The State of Ohio recognized the brothers for their invention; Orville accepted the citation in the Ohio General Assembly. A Dayton elementary school was named for Orville, and a life-size profile of the brothers on a bronze tablet was placed in Wilbur Wright High School, Dayton. After the Ohio Federation of Women's Clubs honored Orville, he dashed to St. Louis to accept an award by the American Society of Mechanical Engineers. Orville received the first Daniel Guggenheim Medal April 8, 1930. The base stone of the monument at Le Mans, honoring Wilbur's first flight there, was laid in November. Orville was selected as one of the thirteen most prominent inventors and scientists in the United States and honored at the opening of the new patent office in Washington in 1932. At universities and colleges around the country a rainbow of honorary doctoral hoods was placed on Orville's shoulders.

The Hudson Motor Car Company presented Orville a new Essex Terraplane automobile, which had been christened by Amelia Earhart. The car was named Terraplane because it was the first to incorporate principles of airplane construction in automobile manufacture.

Orville flew better than he drove. He drove over the lawn of his mansion rather than proceed slowly up the winding driveway. Sometimes he drove across vacant lots rather than waste time going around the block.

Every morning when he was not being an honored guest at some function, he drove his car from Hawthorn Hill to his laboratory. One day when he was on his way home, his car was involved in an accident at the corner of Second and Broadway. Early next morning Police Sergeant Paul Price found that Orville Wright had made an appointment to see him.

When Orville arrived at the police station, Price found him to be modest, courteous and soft-spoken. "I am worried about the accident I had yesterday," Orville told Price. "I should hate it very much if the driver of the other car would sue me. I just wouldn't want to be involved in that kind of suit."

Sergeant Price sent for the accident report and read it. "Why, Mr. Wright, you do not need to be worried at all. I see here that the driver of the other car ran a red light and has been cited. There isn't a thing for you to be worried about."

"But even so," Orville said, "he might sue me for damages to his car."

"I cannot see that there is the least possibility of that. But if such a thing should happen, your insurance company will protect you."

Orville shook his head. "That's just it. I don't have any insurance."

"You mean to say you don't have auto insurance?"

"No. I don't believe in insurance."

"You don't believe in—you mean you don't have fire insurance on your house or office?"

Orville shook his head.

Wright Brothers National Memorial, Kitty Hawk, North Carolina

"How about life insurance?"

"Don't believe in it. I never had any."

Ever after that, Price said, he and the Dayton and Oakwood police said a prayer every time they knew Mr. Wright was on the streets in his car.

On top of Kill Devil Hill at Kitty Hawk, soaring sixty feet into the skies, stands a gray granite shaft shaped like a pylon. Wing-like rays etched into the sides make the shaft look as if it is about to fly. On the face of the shaft are four words:

Wilbur
Wright

Orville
Wright

The inscription on the base of the memorial reads:

IN COMMEMORATION OF THE CONQUEST OF AIR BY THE
BROTHERS WILBUR AND ORVILLE WRIGHT. CONCEIVED BY
GENIUS. ACHIEVED BY DAUNTLESS RESOLUTION AND
UNCONQUERABLE FAITH.

Orville returned to Kitty Hawk November 19, 1932, to be present when the Wright Brothers National Memorial was dedicated. Ruth Nichols unveiled the memorial, which was accepted for the United States by Secretary of War Patrick J. Hurley. Orville accepted the memorial on behalf of his brother Wilbur and himself.

Lindsay C. Warren, Congressman from North Carolina, J. C. B. Eringhaus, governor-elect of North Carolina, and Secretary Hurley spoke. President Herbert Hoover sent a letter of congratulations.

Orville Wright is the only man to live to see a United States National Memorial dedicated to himself.

On the thirtieth anniversary of the first flight, 1933, Orville attended the dedication of the aviation section of the Franklin Institute in Philadelphia, which earlier in the year had awarded him the Franklin Medal. On the same day as part of numerous nation-wide celebrations a delegation placed a laurel wreath on the Wright exhibit at the National Museum as a tribute from the women of America. The wreath was accepted by Dr. Charles G. Abbot, Secretary of the Smithsonian Institution.

Dr. Abbot announced on December 18 that he had appointed a committee of three to investigate the continuing controversy between the Institution and Orville, who stated that he was pleased to learn that the Smithsonian was willing to accept the proposal he made in 1925. He heartily approved Abbot's appointment of Charles A. Lindbergh to the committee. Lindbergh spent some time with Orville but did not understand what Orville wanted. Lindbergh recommended to Abbot that a disinterested writer should prepare an account of all of the controversy and that it should be published by the Smithsonian. When Abbot asked Orville if this would be satisfactory to him, he said it would not. Once again Abbot wrote an article on the controversy and sent it for approval to Orville, and once again Orville did not approve.

Henry Ford, who was building Greenfield Village at Dearborn, Michigan, wanted the old Wright home at 7 Hawthorn Street and the West Third Street Bicycle Shop for his Village. He inquired of Orville whether he could buy them. The home had been in Katharine Wright's name and she had sold it before her death in 1929 to Charlotte Jones, the Wright laundress. One day in 1936 Charlotte Jones answered a knock on the door at 7 Hawthorn Street. Standing on the porch were Mr. Orville and a thin man Mr. Orville introduced as Henry Ford.

The two men sat on the sofa facing her. "Mr. Ford is building a museum and village at Dearborn, Michigan. He would like to move this house and the shed behind it and the bicycle shop to the village. He would like to buy the furnishings along with the house. Now Charlotte, will you sell to Mr. Ford?"

"Well, Mr. Orville, if you say so, I will."

Negotiations for the purchase of both properties were completed before the people of Dayton knew it had happened. When they learned the news in 1936, it was too late to stop the loss of the properties. Old-timers in Dayton still say, "Henry Ford is a robber."

Orville shows Henry Ford the bandsaw used by the brothers. It is in the bicycle shop in Greenfield Village, Dearborn, Michigan.

The buildings were torn down and removed to Dearborn. The dedication took place on the seventy-first anniversary of Wilbur's birth, April 16, 1938. Orville and his family attended as did Colonel Frank P. Lahm, Orville's first Army passenger, Walter R. Brookins, Orville's first American civilian student, and Griffith Brewer.

Not all the furniture from the 7 Hawthorn Street home went along on the moving van to Dearborn. Over the years Charlotte Jones kept finding bits of this and that which had been in the old home. Every time this happened, someone from Dearborn came down, inspected the piece, and then asked Orville whether he remembered it. Orville always answered, "Oh, yes, that used to be in our house."

And so Charlotte added to her bank account as the years passed. One day she found an old chair. Yes, Orville remembered the chair. "How much do you want?" asked the man from Dearborn.

"One hundred dollars."

"Here is a check for five hundred dollars," he said. "This is to cover anything else you find."

Charlotte never found another thing.

Henry Ford did not take this chair to Greenfield Village because Orville took his favorite chair to his library at Hawthorn Hill, where it has remained.

PIONEER FLYERS WHO WERE TRAINED AT WRIGHT BROTHERS FIELD

MAJ. GEN. HENRY H. ARNOLD, U. S. A.
APPOINTED CHIEF OF AIR CORPS. U. S. ARMY, 1938.

BRIG. GEN. FRANK LAHM. U. S. A.
FIRST MILITARY MAN TO GO UP IN AN AEROPLANE.
ONE OF FIRST TWO MILITARY PILOTS IN AMERICA.

CAPT. JOHN RODGERS. U. S. N.
FIRST ATTEMPTED FLIGHT FROM UNITED STATES TO HONOLULU. 1925.
LONGEST MADE BY SEAPLANE TO THAT TIME – DISTANCE 1840 MILES.

CAPT. KENNETH WHITING. U. S. N.
U. S. NAVY FLYER. FIRST MAN TO BE SHOT FROM
TORPEDO TUBE OF SUBMARINE.

CAPT. A. ROY BROWN
CANADIAN ROYAL AIR FORCE. HOLDER OF A DISTINGUISHED
RECORD FOR OUTSTANDING MILITARY ACCOMPLISHMENT.

COL. CHARLES DeF. CHANDLER. U. S. A.
AIR CORPS OFFICER AND FORMER CHIEF OF LIGHTER–
THAN–AIR–DIVISION.

COL. THOMAS DeW. MILLING. U. S. A.
ONE OF THE FIRST MILITARY PILOTS.

GRIFFITH BREWER
FIRST ENGLISHMAN TO FLY IN AN AEROPLANE.

CAL. P. RODGERS
FIRST TRANSCONTINENTAL FLIGHT FROM EAST
TO WEST.

ROBERT G. FOWLER
FIRST TRANSCONTINENTAL FLIGHT FROM WEST
TO EAST.

WALTER BROOKINS · RALPH JOHNSTONE · ARCH HOXSEY · DUVAL LACHAPPELLE · A. L. WELSH · FRANK T. COFFYN
F. O. PARMELEE · J. C. TURPIN · HOWARD GILL · L. W. BONNEY · O. A. BRINDLEY · J. C. HENNING · HAROLD H. BROWN
R. J. ARMOR · HARRY N. ATWOOD · H. V. HILLS · LOUIE MITCHELL · O. G. SIMMONS · C. L. WEBSTER · ALBERT ELTON
ANDREW DREW · A. A. MERRILL · PHILIP W. PAGE · GEORGE A. GRAY · C. COUTURIER · WILFRED STEVENS · ARCH FREEMAN
J. G. KLOCKLER · FARNUM T. FISH · F. J. SOUTHARD · GROVER C. BERGDOLL · CHARLES WALD · WILLIAM KABITZKE
M. R. PRIEST · JOHN A. BIXLER · BERNARD L. WHELAN · HOWARD M. RINEHART · A. A. BRESSMAN · M. T. SCHERMERHORN
R. M. WRIGHT · W. E. BOWERSOX · L. E. BROWN · A. B. GAINES. JR. · C. J. PETERSON · L. E. NORMAN · C. E. UTTER · C. A. TERRELL
MRS. RICHBERG HORNSBY · C. LAQ. DAY · MARJORIE STINSON · C. ANDO · FRANK KITAMURA · O. A. DANIELSON · LYLE H. SCOTT
FERDINAND EGGENA · ROBERT E. LEE · ROSE DOUGAN · J. M. ALEXANDER · J. A. McRAE · GOROKU MORO · VERNE CARTER
E. P. BECKWITH · T. D. PEMBERTON · B. B. LEWIS · MAURICE COOMBS · GEORGE H. SIMPSON · GORDON F. ROSS · K. G. MACDONALD
PERCY E. BEASLEY · K. F. SAUNDERS · M. B. GALBRAITH · W. J. SUSSAN · C. J. CREERY · JOHN GALPIN · BASIL D. HOBBS · JAMES L. GORDON
EDWARD A. STINSON · M. C. DUBUC · J. A. SHAW · P. S. KENNEDY · LLOYD S. BREADNER · W. H. CHISAM · ROBERT McC. WEIR · G. A. MAGOR
N. A. MAGOR · J. R. BIBBY · G. S. HARROWER · GEORGE BREADNER · C. E. NEIDIG · A. W. BRIGGS · H. B. EVANS · A. C. HARLAND · HARLEY SMITH
J. C. WATSON · S. T. EDWARDS · HARRY SWAN · A. G. WOODWARD · L. B. AULT · A. Y. WILKS · J. C. SIMPSON · PAUL GADBOIS
C. McNICOLL · W. E. ROBINSON · M. S. BEAL · C. G. BRONSON · W. E. ORCHARD · J. A. HARMAN · T. C. WILKINSON · J. G. IRELAND

Bergdoll's name is inscribed on this plaque five lines below the aeroplane. The plaque is on the Wilbur and Orville Memorial, Wright Brothers Hill near Simms Station, Dayton.

When plans were drawn for the Wilbur and Orville Wright Memorial in Dayton, the committee decided to list the names on the memorial of all pioneer flyers who were taught to fly by the Wright brothers. When Orville looked over the list, he said, "You have omitted a name."

"We have omitted the name of Grover Cleveland Bergdoll," said the committee chairman. "And we did it on purpose. He is not worthy of inclusion."

Grover Cleveland Bergdoll had begun his training at the Wright School in 1912. He became one of the best flyers in the country. A wealthy young man, he bought an aeroplane of his own and gave exhibitions at his estate near Philadelphia. In 1917 when the United States entered World War I, Bergdoll refused to register for the draft. He became notorious as a draft-dodger.

Orville handed the list to the committee chairman. "If Bergdoll's name is not on the monument," said he, "then you don't need to put mine on either."

Bergdoll's name is on the memorial, which stands on Wright Brothers Hill, overlooking the site of Simms Station near Dayton. The memorial was dedicated August 19, 1940, Orville Wright's birthday and National Aviation Day. At the dedication the Civil Aeronautics Board presented Orville with Honorary Pilot's Certificate No. 1.

When in 1940 Franklin Roosevelt campaigned for his third election to the presidency, he stopped in Dayton on October 13. A thirty-five car motorcade met his train at Union Station. In the presidential open touring car Orville sat between Roosevelt and James M. Cox, former governor of Ohio. The motorcade drove over thirty-four miles of Dayton streets, lined with cheering people. It stopped for a few moments at the Wright Memorial and then set out for Cox's home, Trailsend, where luncheon was waiting. As they drove through Oakwood, Orville tapped the driver on the shoulder. "Just pull over to the curb," he said. "I live about a mile from here and I'll walk home."

The car stopped. Orville climbed out, shook hands with the President and walked off. Orville's place at the Cox luncheon table was empty and his family was horrified that he had walked out on the President of the United States. Orville was a Republican.

One Republican between two Democrats: President Roosevelt, Orville Wright, and James M. Cox, 1940.

Dr. Charles Abbot in 1942 tried once more to resolve the controversy between the Smithsonian and Orville. Abbot wrote to Orville, asking him to set down in writing what he wished the Smithsonian to do. Orville sent him a long list of differences which had existed since 1914. Abbot was unable to check the differences but because he realized that the fortieth anniversary of the Kitty Hawk was approaching and that Orville was seventy years of age, and that if he died before the controversy was settled, the Smithsonian would suffer great prejudice. He wrote an article yielding to everything Orville asked without making any reservations.

Abbot's paper reviewed the controversy, listed the changes made at Hammondsport, expressed regrets for the erroneous statements made in the past by the Institution, and concluded:

The flights of the Langley aerodrome at Hammondsport in 1914, having been made long after flying had become a common art, and with changes of the machine indicated by Dr. Wright's comparison, did not warrant the statement published by the Smithsonian Institution that these tests proved the Langley machine of 1903 was capable of sustained flight carrying a man.

If the publication of this paper should clear the way for Dr. Wright to bring back to America the Kitty Hawk machine to which all the world awards first place, it will be a source of profound and enduring gratification to his countrymen everywhere. Should he decide to deposit the plane in the United States National Museum, it would be given the highest place of honor, which is its due.

Abbot sent the article to Orville, stating that he would publish it with such changes as Orville wished, but only if he would prefix to it a statement of his own that he was satisfied.

If the paper would be printed in the next Smithsonian *Annual Report,* Orville said, he would be satisfied. He wrote in part to Dr. Abbot, "I can well understand the difficult position you found yourself in when you took over the administration of the Institution at a time when it had on its hands an embarrassing controversy for which you were not responsible; so I appreciate the more your effort to correct the record of the tests at Hammondsport which brought on this controversy."

Abbot's article appeared in the No. 1 position in the "General Appendix" to the 1942 *Annual Report.*

Early in January 1943 Dayton postmaster, Clarence N. Greer, working with the Dayton Chamber of Commerce, began a campaign to have the United States Post Office issue a commemorative stamp to honor the Wrights while Orville was living.

Charles J. Wood designed a stamp. Using a photograph of the first flight, the artist printed in the left panel "Orville Wright" and in the right "Wilbur Wright." In the sky above the plane Wood lettered

FIRST FLIGHT OF THE FIRST PLANE
CAPABLE OF SUSTAINED FREE FLIGHT

Wood and Greer took the drawing to Orville for his approval. After he studied it, he said, "I have some objections."

"Will you tell me what they are?"

"I don't care to say."

Wood questioned him further. Finally Orville said, "I object to the wording over the plane."

"But it is the truth, isn't it, Mr. Wright?"

"Yes, it is true, but I think the Post Office is not the proper group to make that statement."

Wood, realizing that he had unintentionally brought the old controversy to Orville's mind, said, "Suppose we change the wording to this:

FIRST FLIGHT OF A PLANE
CAPABLE OF SUSTAINED FREE FLIGHT

"That statement now expresses with exactness and simplicity the full significance which should characterize the stamp," Orville said. "If you change the wording, I agree. However, I object to the use of the Wright name on the stamp while one of us lives."

Wood then suggested that the left panel should say "Dayton" and the right one "Kitty Hawk."

"Yes. In all honors connected with the first flight, the names of both towns should be used."

Although Charles F. Kettering and former governor James M. Cox used their influence in Washington, the United States Post Office did not issue the stamp because in those days the office issued ten commemorative stamps a year and its 1943 quota had been filled.

Orville Wright to President Franklin D. Roosevelt

Dayton, October 16, 1943

I am greatly honored to have from the President of the United States an invitation to come to Washington on December 17th. Under the circumstances I have had no difficulty in securing releases from other obligations for that day.

I am sorry that inabilities as a speaker compel me to decline any speaking part in a program.

I believe it is time that we should be giving consideration to postwar matters; and I hope that when the times comes our people will be willing to make such sacrifices as are necessary to secure a lasting peace.

Orville Wright to President Franklin D. Roosevelt

Dayton, November 16, 1943

I think your suggestion of having the announcement of the future return of the Kitty Hawk plane made at the dinner on December 17th a good one. I had intended that the announcement be made as soon as the Smithsonian report appeared, but I now shall have it withheld for the dinner. If you personally were to make the announcement it would be particularly pleasing to me.

The dinner did not go off very well. In the first place, when Orville arrived in Washington, he learned that Roosevelt was out of the country and would not be able to attend the dinner or make the announcement. Then just as he was dressing to go to the White House, he was informed that he was expected to present the Collier Trophy to General Henry H. Arnold. Orville's temper rose. Although he was a friend of Hap Arnold, he already said he would not speak and he made it plain that he did not intend to.

Newsreel cameramen photographed the proceedings and the speeches were broadcast over national radio. At the conclusion of his speech, Secretary of Commerce Jesse Jones announced that Orville Wright would now present the Collier Trophy.

With the camera lights on him and the radio microphone held to his face, Orville stood up. He handed the trophy to Hap Arnold and sat down. Not one syllable did he say. Arnold covered the embarrassing silence by saying, "There is no one in the world from whom I would rather receive the Collier Trophy than Orville Wright."

Although Orville rarely spoke in public, in private he was more talkative. Once one of his grandnephews invited him to spend an evening at his fraternity house at Miami University. Orville told the boys gathered around him one story after another about the days of the early flights. One boy said that there was something in his aviation textbook that he did not understand. "Let me see it," said Orville. He read the statement. "Oh, that's wrong." He pulled out a pen, wrote a correction in the margin, and signed it "O.W."

Ten children in Orville's neighborhood walked up onto the porch of Hawthorn Hill one Halloween and set up a great racket on the windows with their tick-tacks, made of old wooden spools with notched edges. After a while Orville opened the front door, looked at the group, pointed to two children and said, "You two come inside. The rest of you wait outside." He asked the two children to sit down, quizzed them about their school, where they lived, and what their fathers did. Then he took out from his pocket a fifty-cent piece. "You take everybody out there to the drug store and buy them an ice cream cone," he said. And off they went. Twenty years later, according to Frederick C. Smith and Anthony Bailey, Orville gave each Halloween visitor a silver dollar.

Dr. Theodore Lilly, a Dayton dentist, met Orville when Katharine gave a party for the local Oberlin alumni. One morning shortly afterwards a woman called his office. "This is Mabel Beck, Mr. Wright's secretary. He would like to have an appointment with Dr. Lilly."

Lilly laughed. "Oh, sure it is," he said. "Now why would Orville Wright want to do that?"

"Because he needs some dental work," said Mabel sharply. "He has heard that you are a good dentist."

Lilly, still thinking somebody was playing a joke, said, "I was hoping nobody would find that out." When Mabel, who had a razor tongue, gave him a piece of her mind, Lilly made the appointment.

When Orville arrived for his appointment Dr. Lilly apologized. Orville laughed and said Mabel was always difficult and had no sense of humor. In all the years Orville went to Dr. Lilly he was late only once. Then he apologized. He had found that some chipmunks had plugged the drain tile and in cleaning the drain, he had let the time get away from him.

Orville did all the household repairs. Under the bathroom floors he installed metal shields so that if ever the bathroom should leak or somebody let the tub run over, the water would not damage the ceilings below.

He used rain water for the cold and hot bath water. First the water was collected from the spouting on the roof into an underground cistern. From there he had it piped through a special filter he designed to clear out impurities. The water was piped into a second cistern and then into the house. Once Carrie Grumbach told Jay R. Petree, one of Orville's nephews, that the water heater had not been working for a long time. Jay offered to repair it. "Oh, no," said Carrie. "Mr. Orville would have a fit. He has promised to fix it himself."

Edward Deeds, Charles F. Kettering, and Orville often spent evenings together. After dining, they sat around spinning yarns. One of the men said he had heard it was a good idea to lie down after a heavy meal. Deeds and Kettering discussed the matter for some time, finally deciding that it is not a good idea because blood circulation needed for digestion slows down during a nap. Then Orville, who had said nothing, remarked, "If what you fellows say is true, there must be a lot of sick dogs in this world."

Colonel Edward A. Deeds, Henry Ford, and Charles F. Kettering join Orville Wright for the 35th anniversary program at the NCR auditorium December 17, 1938.

Always a perfectionist, Orville inspected every inch of the dining room table before a dinner party. Sometimes he said to William Lewis, his butler, "William, the tablecloth is one inch longer on this side than on the other side. Please reset the table and change the cloth." William used a yardstick to measure the distances between the plates and the distance the knives and forks and spoons were from each other. During dinner whenever Orville tapped his water goblet, William knew that a guest had taken a drink. The tapping was the signal for William to refill the goblet because Orville insisted that the water level in each goblet on the table must be the same.

Orville required that all the rugs in Hawthorn Hill had to be lined up with the cracks in the oak or walnut floors. Whenever Orville walked on any rug, Carrie or William had to run the vacuum sweeper, not because of the dirt but because he did not like to see footprints on his rugs. Sometimes when Orville was not looking, William used a broom to turn back the nap on the rugs. Orville also required Carrie and William to walk on the edges of the rugs to protect and save the middle of the rugs.

During one of his visits to Hawthorn Hill, Fred Kelly, the biographer, asked Orville, "Can you remember what happened the Monday morning when you started out for your first day in kindergarten?"

Orville sat quiet for a long time. "I'm not sure," he answered, "that it was on a Monday."

Orville bought an IBM typewriter. As soon as it was delivered, he began to take it apart to see how it was made. Soon the machine was lying in pieces in a bushel basket. Then he tried to put it together again. Said he to Mabel Beck, "Call the IBM for a repairman."

When IBM arrived, the repairman said, "I can repair typewriters but I can't assemble them." He took the basket of pieces away. Several days later IBM sent Orville a new typewriter with a sign that said, "Think."

One fall morning Orville walked up the driveway of the Loren M. Berry house and

Orville's rolltop desk and drafting table in his North Broadway Street office, Dayton

Orville's neighbor, Loren M. Berry [1888-1980], sits under the old oak tree where Orville's picnic bench used to be.

started to the backyard. Loren Berry greeted him. "I'm glad to meet you," said Orville. "I wanted to look at the big oak tree in back of your house. It was our favorite picnic place for many years."

They walked to the oak tree. "Our family used to come here on Sundays when we owned this land. What's become of the picnic bench?"

"You mean that old bench around the tree?"

"Yes. I built that bench in my basement up the hill and after I built it, I had to take it apart to get it out of the basement. I brought it down here in pieces and put it together again. We always liked the bench because it was wide and we could sit on it. If it rained, we didn't get wet because the tree has very thick foliage."

Berry said, "Mr. Wright, I told the contractor to remove that bench because it had been damaged by a truck that came here when we were building the house and dumped a lot of brick on the bench. I didn't even know you had owned this property. If I had known you had built that bench, I would have had a very fine plaque—bronze or something better—put up and it would always be there for all time and I'm certainly sorry I didn't know that." Orville said, "Oh, well, it wasn't a very good bench anyhow."

Loren M. Berry did place a bronze plaque on the tree. It reads:

ORVILLE WRIGHT TOLD LOREN BERRY
THAT AROUND THIS OLD OAK TREE
HE HAD BUILT A BENCH TO WHICH
HE CAME MANY TIMES TO PICNIC

In 1929 five-year-old Elizabeth Ann, Loren Berry's daughter, asked her father if she and her friend Barbara Smiley could go around the neighborhood on Halloween night. He drove the girls to Harman Avenue. The two girls walked up the long winding driveway to the white house with the pillars. They rang the doorbell twice. A balding, mustached man opened the door, looked at the costumed visitors and invited them in and asked them to sit down. "Would you like a glass of milk and some cookies?" They would, and he procured them. He asked them about their school and what their fathers did. "Did one of your fathers drive you here?"

"Mine did," Elizabeth said. "He wouldn't let us go by ourselves." The host stood up. "I think he's honking the horn." He walked to the door and listened. "Yes, somebody's honking. I think you better leave. But come back again."

On December 17 of that year Loren brought home a dozen red roses. Calling Elizabeth to him, he said, "I will drive you to the house of your friend on the hill. I want you to give him these roses. This is a special anniversary."

Every December thereafter Elizabeth Berry, even after she had grown up and married, sent a bouquet of red roses to Orville Wright.

Orville Wright to Elizabeth Berry Fox

December, 1947

The roses on the seventeenth were beautiful. I think I have never seen more gorgeous ones. Wishing you both and little Stuart Charles a Merry Christmas and Happy New Year.

Harold H. Haas to The Journal Herald

Dayton, September 24, 1978

In the summer of 1946, when the first Lockheed *Constellation* to visit Dayton arrived at the Dayton airport there was quite a ceremony made of it. A flight was made from the airport on which several prominent Dayton officials and particularly Orville Wright were passengers. Orville, of course was the honored guest. It was announced that he would fly the plane after it became airborne.

At that time I was working in the plant protection department at Aeroproducts division, General Motors, across the street from the airport. I was on loan to the airport for guard duty that day. I happened to be stationed at the foot of the portable stairs leading to the plane. As Orville Wright started up these stairs, he stopped and shook hands with me. He said, "I want to talk to you when we get back."

I had known Mr. Wright for years because my father, Walter E. Haas, had made patterns for some of the castings used in the first Wright plane engine.

After the *Constellation* returned to the airport, Mr. Wright came down the stairs and stopped to talk to me. "Harold, I want you to be the first to know that I am the only person in the world that has lived through and has been directly associated with a principle of mechanics from its inception to its obsolescence. Wilbur and I produced and flew the first airplane powered by a conventional internal combustion engine, and I have just had the pleasure of flying one of the last models of a commercial plane that is powered by an internal combustion engine. All future commercial airline planes will be powered by some type of a jet propulsion engine. I am more proud of this unique experience than any other that has ever happened to me."

On December 9, 1947, Orville asked two men to come to the laboratory to help him clean all the junk out of the attic. The men handed down box after box to Orville, who stood below. He inspected each box and decided whether it should be saved or thrown

Orville kept an interest with advances in aviation. He and Colonel Edward Deeds inspect the German rocket-propelled interceptor used in World War II.

away. One of the men handed him an old metal cover for a Smith Premier typewriter. "That's junk," said Orville and started to put it on the discard pile. Something in the box rattled. He shook it. When he opened the box he found inside the balances for the original wind tunnel, the ones he had his brother Lorin carry by hand in 1916 from 1127 West Third Street to the laboratory at 15 North Broadway because he didn't trust the movers. Orville had never found the balances in the laboratory and accused Lorin of losing them. The balances were somewhat rusty, having been through the 1913 flood, part of one dial was missing as were two other small pieces, but the rest were in good condition. Said Orville, "I'll just fix these up."

The balances had been made in 1901. Although they were crudely made out of old bits of junk from around the bicycle shop, they were surprisingly accurate. George W. Lewis, head of the National Advisory Committee for Aeronautics, said in London at the twenty-seventh annual Wilbur Wright Memorial Lecture, "Using the average flight measurements on their 1902 glider and the wind tunnel results on models, the Wright brothers calculated that the unit drag of a square flat plate of large area was equal to the square of the speed in miles per hour multiplied by the coefficient 0.000330. Much subsequent research has led to the general adoption of the figure of 0.000328 (only two hundred thousandths difference), truly astonishing agreement with the value reached by the Wright brothers averaging a series of results obtained in flight measurements in 1902."

It was these balances and the wind tunnel that enabled the brothers in 1901 to solve the mystery of air pressure forces on wing surfaces, a problem that had eluded all their predecessors.

General Henry H. Arnold, retiring Air Force Commander, had recommended early in 1946 that Orville should be given the Order of Merit for distinguished service with the National Advisory Committee for Aeronautics during World War II. President Harry

S. Truman signed the certificate April 1, 1947. The award ceremony was planned for January 15, 1948.

Orville was two minutes late for an appointment with Edward Deeds at the National Cash Register Company on the morning of October 10, 1947. He ran up the steps of the headquarters building and fell. At Miami Valley Hospital his physician issued a statement that Orville Wright was under observation for chest pain. After four days in the hospital he returned to Hawthorn Hill. He wrote to Frank P. Lahm on November 12, "If they had not taken my clothes away from me, probably I would have been out of the hospital a day or two sooner. I am now almost back to normal health and am driving a car and have been coming to the office every day for several weeks."

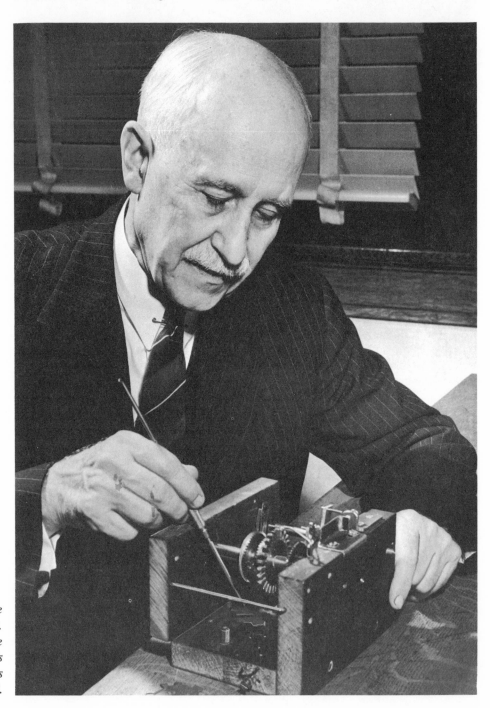

Do you remember the time Orville oiled Mrs. Sines's sewing machine with water? Here he is sixty-seven years later, 1943.

Sec. of Commerce Jesse Jones reads Roosevelt's message at the White House dinner to commemorate the 40th anniversary of powered flight, 1943.

Orville's former student pilot, General Hap Arnold, awakens Orville and accepts the Collier Trophy at the dinner.

He wrote on January 12, 1948, that his health would not allow him to go to Washington to receive the Award of Merit. The award committee decided that at some later date a government official, probably General Carl Spaatz, would present the Award in Dayton.

Orville Wright to the General Electric Company

Dayton, January 24, 1948

I have a 10-H.P. General Electric motor that I would like to sell. This motor was purchased in 1917 for use in my laboratory on a wind tunnel. I would estimate it had about 200 hours of running. I am quoting from the invoice of the General Electric Company of Schenectady of February 10, 1917; 1 KT-322-10-720/685-220 V 60 cyc "B" motor 1186788 DRL-1891926 amps 28.

Would you be interested in purchasing it or do you know of an interested purchaser?

After breakfast two mornings later he discovered that the front door bell and the bells from various rooms to a board in the kitchen were not working properly. He went to work on the bells. Up and down the steps he went, in and out of the house. When time came to go to the laboratory, he told Carrie, the housekeeper, that he would finish the bells when he came home.

Not long after he arrived at the laboratory and began to work at his desk, he felt ill. His secretary Mabel Beck called his physician, who ordered an ambulance to take Orville to Miami Valley Hospital. Three days later on January 30, 1948, he died in his sleep. He was in his seventy-seventh year. His physician issued a statement that death was due to arteriosclerosis and lung congestion.

The day of Orville's funeral, February 2, all municipal and county offices and all schools in Dayton and Oakwood, public and parochial, and the University of Dayton closed at noon. Services were held at the First Baptist Church at 2:30 p.m. Dr. Charles L. Seasholes, the only Dayton clergyman whom Orville approved, conducted the services.

Forty-five of the fifty honorary pallbearers attended the services, including nine generals and four admirals. Some of the guests were Dr. Alexander Wetmore, secretary of the Smithsonian; Brigadier General T. D. Milling; Dr. Hugh Dryden, director of the National Advisory Committee on Aeronautics; John F. Victory, executive secretary, NACA; S. P. Johnson, director of the Institute of Aeronautical Sciences; John Allison, assistant Secretary of Commerce for Air; Major General L. C. Craigie, former chief engineer at Wright Field, representing the USAF in Washington; Rear Admiral T. C. Lennquest, representing NACA; R. W. Hazen, director of Allison Division, General Motors Corporation.

Others included General Carl Spaatz, Commander of the United States Air Force; Vice Admiral John H. Price, Department Chief, Naval Operations for Air; General Henry H. Arnold, Air Force Chief Ret.; General J. T. McNarney, Commander, Air Materiel Command Headquarters at Wright Field; Admiral Ernest King, Chief of Naval Operations, Ret.; Lieutenant General Nathan F. Twining, former AMC head, Commander of Ground, Air, and Sea Forces in the Alaska Theater; General Frank P. Lahm, Ret.; General William F. Gillmore, Ret.; Rear Admiral F. W. Pennoyer, Navy liason, Wright Field; General Benjamin D. Foulois, Ret.

Also, Dr. Vannever Bush, Carnegie Institute; Earl Findley, publisher, *U. S. Air Services* magazine; Henry Ford, II; Dr. C. J. Honsaker, chairman NACA; Sir Paul Johnson, director of the Institute of Aeronautical Sciences; Glenn L. Martin; F. W.

James M. Cox speaks at the dedication of the Wilbur and Orville Memorial, August 19, 1940. It was Orville's 69th birthday and he was there. Leontine Jameson and Marianne Miller, grandnieces of Orville, unveiled the monument.

Reichelderfer, Chief, U. S. Weather Bureau; Gordon S. Rentschler, board chairman, National City Bank of New York; Dr. George W. Lewis and William E. Butterwood of NACA.

Seven of Orville's flying pupils attended: Walter Brookins, Griffith Brewer, Robert Fowler, Major Lester D. Gardner, R. M. Wright, Grover Loening, and Bernard Whalen. Others were mechanic Harvey Geyer, biographer Fred C. Kelly, Frank McCormick, Charles E. Taylor, and Captain William J. Tate of Kitty Hawk.

Prominent Daytonians at the services included Charles F. Kettering, Colonel Edward A. Deeds, Governor James M. Cox, Dr. A. R. Brower, Mayor Edward Breen, William A. Chryst, Dr. Frank Henry, Paul Ackerman, Lewis B. Rock and Walter H. J. Behm.

Six months after Orville's death the 1903 Wright Flyer comes home. The crated machine is lowered into a van from the U.S.S. Palau at Bayonne, New Jersey.

The van carrying the crated Flyer arrives at the Smithsonian Institution, November 22, 1948. The sign on the van reads "The Kitty Hawk."

A battle formation of four Air Force planes—with the fifth missing—flew over the funeral cortege as it drove the length of Main Street to Woodland Cemetery. As the planes flew over the grave site, they dipped their wings in a farewell salute.

Orville's estate was valued at $1,023,923. In his will signed June 21, 1937, he left $300,000 to Oberlin College and bequests to be paid during the lifetime of his brother Lorin; Reuchlin's widow; Mabel Beck; Ed Sines, his boyhood friend; Charles Taylor, Carrie Kaylor Grumbach, Charlotte Jones. He bequeathed his Hawthorn Hill furnishings and the Lambert Island property and furnishings to his nieces and nephews. The residue of his estate was divided among his nieces and nephews and grandnieces and grandnephews.

Orville left his medals to the trustees of the Dayton Art Institute. He left the wind tunnel balances to the Franklin Institute and bequests to Berea College and Earlham College.

He bequeathed the Kitty Hawk flyer to the Science Museum of South Kensington, London—"Unless before my decease I personally in writing have asked its withdrawal from that Museum." In a letter to the Museum director in 1943 he had asked that the aeroplane be returned at a time when transportation was safe. In an unfinished will he was working on at the time of his death he again indicated the aeroplane should be returned and asked that it be placed in the National Air Museum of the Smithsonian Institution.

Dr. Herman Shaw, Keeper of the Kensington Museum, accompanied the Kitty Hawk Flyer on the *Mauretania* to New York in November 1948. The aeroplane was stored in three wooden crates. Paul Garber, then a curator and historian of the Smithsonian's National Museum, drove a Navy truck to the docks in New York City. On the side of the truck he had hung a sign, "Operation Homecoming." There he learned that a dock strike forced the *Mauretania* to go to Halifax. Garber arrived in Halifax just as the *Mauretania* was docking, but by that time the strike had closed the Halifax harbor.

Garber explained to the cargo chief that the crates contained a national treasure and he agreed to lower them over the side. Garber secured some pallets, put the crates on them, and pushed them under shelter.

He telephoned Admiral Alfred M. Pride, head of the Bureau of Aeronautics in Washington, and asked if the Navy could pick up the crates. The Admiral sent the *U. S. S. Palau,* a small aircraft carrier. Every day until the carrier arrived on November 14, Garber sat beside the crates on the dock at Halifax. As the carrier dropped anchor, the Canadian ships in the harbor gave a salute of guns and whistles. Garber said in a 1966 *New Yorker* interview that he turned to a man standing next to him and said, "My, that was a nice welcome."

"That wasn't for your ship," said the man. "Princess Elizabeth has just had a baby boy."

After the crates were hoisted aboard the carrier, the ship proceeded to Bayonne, New Jersey, where a Navy truck was waiting. When the Kitty Hawk flyer was ready to be placed on the North Hall facing the entrance, the place of honor was occupied by *The Spirit of St. Louis.* Garber called Lindbergh and asked if he minded if the Smithsonian moved his plane to the rear. "Among the honors I have received," said Lindbergh, "there are few I value as highly as sharing space with the Kitty Hawk Flyer."

Visitors to The National Air and Space Museum today see a mustached manikin in a dark pin-striped suit lying propped on his elbows near the center of the lower wing of the Kitty Hawk machine. Garber placed the manikin of Orville in the aeroplane because it makes it easy for visitors to tell which direction the machine flew. A professional sculptor made the head. The suit and shirt belonged to Garber, and he bought a new pair of shoes but they looked so new that Garber had one of the museum staff wear them a day to scuff them.

When he was dressing the manikin, Garber realized he didn't have a high stiff collar. He went to several stores in Washington and Baltimore but none carried stiff collars. Then one evening he saw a small haberdashery near the Willard Hotel. Garber asked the proprietor if he had any stiff collars. "What size?"

"Fifteen and a half."

The haberdasher sold him two collars for fifty cents each. Garber asked, "How do you happen to have these?"

"President Hoover used to buy them from me, and I keep them in stock because he still comes in on occasion."

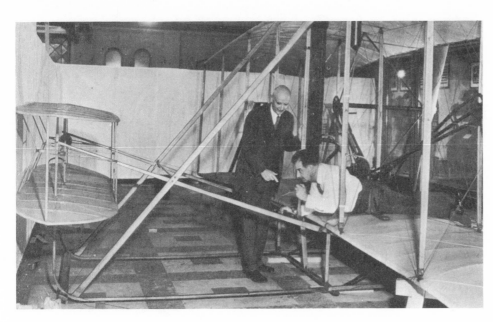

"That's how Orville flew in 1903," says engineer J. Roy Waite, who knew Orville and flew the Wright Model B, to Paul Garber, curator and historian of the Smithsonian.

On the forty-fifth anniversary of the 1903 flight the Wright brothers 1903 aeroplane was formally presented to the Smithsonian Institution.

Dr. Alexander Wetmore, Secretary of the Institution, opened the ceremonies. Greetings came from Chief Justice of the United States and Chancellor of the Institution Fred M. Vinson and from President Harry S. Truman. Milton Wright, son of Lorin Wright, presented the aeroplane to the nation on behalf of the estate of Orville Wright. It was accepted by Chief Justice Vinson. The formal acceptance address was given by Vice President-elect Alben W. Barkley, a Regent of the Smithsonian, and British Ambassador Sir Oliver Franks spoke on "Britain and the Wright Brothers."

Remarks by Milton Wright on Behalf of the Estate of Orville Wright
Washington, December 17, 1948

In the years before 1903 I spent whatever time I was allowed playing about my uncles' bicycle shop. The odor of the glue pot, the spruce shavings on the floor, and the many gadgets whose use I did not understand, were all a great attraction to a small boy. The matter-of-fact way in which my uncles used the gadgets and planed the spruce strips and glued them together into ribs for their flying machine left me with the impression that all bicycle shops did the same thing. It was all very commonplace.

When they took their glider to Kitty Hawk, I thought it must be a fine place to take a vacation, particularly after they sent me such interesting souvenirs as a dried horseshoe crab and bottles containing genuine sea water and sea sand. But I had heard of others taking vacations in equally interesting spots.

They spent many hours making up jointed sheet metal figures for shadow-graph shows with which they entertained all of the children. They made helicopters out of bamboo, paper, corks, and rubber bands and allowed us to run after them when they flew them. There was nothing unusual about that. All uncles probably did the same things. History was being made in their bicycle shop and in their home, but the making was so obscured by the commonplace that I did not recognize it until many years later. The world has come to think of Orville and Wilbur Wright either as demigods whose minds suddenly produced the answer to the problem of flight or as ignorant mechanics who stumbled on the secrets of flight. They were neither. They were normal young men who had an idea and saw a problem and set about to solve that problem. Their ability to select the vital parts of their problem and to discard the unessential was an important factor in their success.

Chief Justice Fred M. Vinson on behalf of the Smithsonian Institution accepts the 1903 Flyer in a ceremony December 17, 1948, forty-five years after the twelve-second flight.

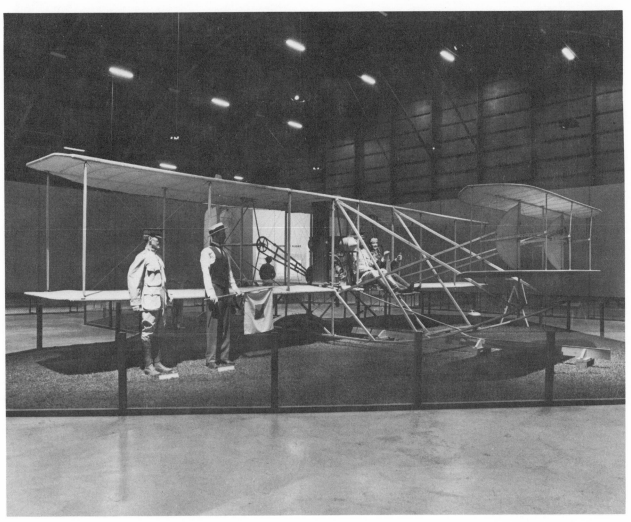

*The modified Model A
aeroplane known as the
1909 Signal Corps Flyer
is the first military
airplane in the world.
This is a reproduction on
exhibit at The United
States Air Force Museum,
Dayton, Ohio.*

I remember their wind tunnel only as a machine that made a lot of noise and was stored in our barn for many years after they finished with it. But the wind tunnel, although inexpensive and crude, was designed with such knowledge of scientific methods that the resulting tables of aerodynamics differ very little from those used today.

When on December 17, 1903, the first flight was made at Kitty Hawk, I was shown a copy of the *Cincinnati Enquirer* which reported that event. I was not impressed by the fact that they had flown because from conversations I had overheard, I knew they had figured out how to do it, and things they figured out usually worked. But I was thrilled by their getting their names in the paper.

Since no one in our family could afford to waste good wood or metal or fabric, it was usual to use parts of old machinery to make up new machines. And so it came as a surprise to me many years later to learn that the first plane had not been dismantled for such purposes—that it was still in existence. It had been carefully packed and put away and it is here today. This important departure from the usual had, I think, a special significance. Wilbur and Orville Wright recognized the fact that their accomplishment was a definite milestone in history and that some day, some place, this first aeroplane would become the symbol of the beginning of an era.

Orville Wright saw clearly that our national capital was the only logical place to set up this symbol. There it could be seen by all and would be forever preserved by our government. The plane, stored in the bicycle shop, had gone through the Dayton flood and was constantly exposed to the threat of fire and other dangers. Since no suitable place seemed to be available in our county, he finally in 1928 carefully restored it in every detail and sent it for safe-keeping to his friends in the British Empire. There it was hung in a place of honor in the Science Museum at South Kensington, London, and was carefully protected by the officials of that institution even through the days of the blitz. After the death of Orville Wright, we found that he had made a request to the Science Museum for the return of the plane to this country. The officials of that institution immediately informed us that they were prepared to return it as he wished. We owe much to them for their cooperation. We also owe much to the present officers of the Smithsonian Institution who made it possible to bring home the plane to stand in its honored place in Washington, as its builders always wished. Even with his dislike of ceremony Orville Wright would have enjoyed this ceremony today.

It was a happy combination of circumstances that, through the seemingly commonplace beginnings, the years of painstaking work to perfect the plane, and later to gain recognition for it, neither of the brothers ever abandoned the vision of this first plane as the symbol of the turning point; the concrete evidence of the fact that what was impossible before in now universally accepted practice.

The aeroplane means many things to many people. To some it may be a vehicle for romantic adventure or simply quick transportation. To others it may be a military weapon or means of relieving suffering. To me it represents the fabric, the glue, the spruce, the sheet metal, and the wire which, put together under commonplace circumstances but with knowledge and skill, gave substance the dreams and fulfillment to hopes.

In the Milestones of Flight Gallery of the National Air and Space Museum stands today the Apollo II command module *Columbia.* A flight of four days, six hours, forty-four minutes, and six seconds in July 1969 carried astronauts Neil Armstrong, Edwin E. Aldrin, Jr., and Michael Collins to the moon, where Neil Armstrong became the first man to step on its surface.

Twenty feet above the space module *Columbia* in the National Air and Space Museum in the very aeroplane in which Orville Wright in 1903 made from the sands of Kitty Hawk the twelve-second flight that sixty-six years later enabled man to land on the Sea of Tranquility and to walk on the moon.

Flight Log

1900	Kitty Hawk, North Carolina
1901	Kitty Hawk, North Carolina
1902	Kitty Hawk, North Carolina
1903	Kitty Hawk, North Carolina
1904	Huffman Prairie, Dayton, Ohio
1905	Huffman Prairie, Dayton, Ohio
1908	Kitty Hawk, North Carolina
1908	Hunaudières Race Course, Le Mans, France
1908	Camp d'Auvours, Le Mans, France
1908	Ft. Myer, Virginia
1909	Pau, France
1909	Centocelle Field, Rome, Italy
1909	Ft. Myer, Virginia
1909	Tempelhof Field, Berlin, Germany
1909	Governors Island, New York
1909	Bornstedt Field, Potsdam, Germany
1909	College Park, Maryland
1910	Montgomery, Alabama
1910	Simms Station, Dayton, Ohio
1911	Kitty Hawk, North Carolina
1911-1918	Simms Station, Dayton, Ohio

Bibliography

Abbot, Charles G. *The 1914 Test of the Langley Aerodrome.* Washington: The Smithsonian Institution, October 24, 1942.

Bryan, C. D. B. *The National Air and Space Museum.* New York: Harry N. Abrams, Inc., 1979.

Dayton Evening Herald. May 30, 1912, p.1.

Dayton Herald, January 30, 1948, p.1.

Cox, James M. *Journey Through My Years.* New York: Simon & Schuster, 1946.

"Dedication of the Wright Brothers Home and Shop in Greenfield Village." Dearborn: The Edison Institute, 1938.

Emerson, Ralph Waldo. "Compensation." *Essays: First Series.*

Foerste, August F. *Geology of Dayton and Vicinity.* Indianapolis: Hollenbeck Press, 1915.

Frost, Robert. "Kitty Hawk." *In the Clearing.* New York: Holt, Rinehart and Winston, 1942.

Garber, Paul E. "The Wright Brothers Contributions to Airplane Design." Speech delivered at the Air Force Museum, Dayton, Ohio, December 14, 1978.

Gibbs-Smith, Charles Harvard. *The Wright Brothers.* London: Her Majesty's Stationery Office, 1963.

Hellman, Geoffrey T. "The Smithsonian Institution." *The New Yorker,* December 10, 1966, Vol. 42, pp. 64-139.

Holy Bible. Revised Standard Version. II Samuel 14:14.

Housman, A.E. "Reveille." *A Shropshire Lad.* New York: Dodd, Mead & Company. 1931.

Kelly, Fred C. *The Wright Brothers.* New York: Harcourt, Brace & Co., 1943.

Langley, Samuel P. "The 'Flying-Machine.' " *McClure's Magazine,* Vol. ix, pp. 647-660.

Loomis, Gilbert J. "Letter to Sidney Strong." March 24, 1960.

McMahon, John R. *The Wright Brothers, Fathers of Flight.* Boston: Little, Brown and Co., Inc.

Milton, John. *Poetical Works.* Douglas Bush, editor. London, New York and Toronto: Oxford University Press. 1969.

Munro, H.H. "The Square Egg." *The Infernal Parliament.* 1924.

New York Times. "Fliers Must Pay Him, Says Wright." Feb. 27, 1914, p. 1.

————. "Curtiss Denies Any Debt to Wrights." March 8, 1914, p. 1.

Roseberry, C.R. *Glenn Curtiss: Pioneer of Flight.* Garden City, N.Y.: Doubleday Company, Inc.. 1972.

Rossetti, Christina. "Uphill." *The New Oxford Book of English Verse.* Helen Gardner, editor. New York and Oxford: Oxford University Press. 1972.

Santayana, George. "O World, Thou Choosest Not." 1894.

Shakespeare, William. *Macbeth,* Act I, vii, 59.

————. *Twelfth Night,* Act V, i, 380.

Shaw, Herbert. "Orville Wright Finds a Historic Relic, Long-Lost." *U.S. Air Services,* Vol. 32, No. 1. January 1947, pp. 17-18.

Smithsonian Institution. "Remarks by Milton Wright, on behalf of the Estate of Orville Wright, in presenting the Kitty Hawk Aeroplane to the United States of America." Press release, Dec. 17, 1948.

Taylor, Charles E. "My Story of the Wright Brothers as told to Robert S. Ball." *Collier's.* Vol. 122. No. 26, pp. 27,68,70. Dec. 25, 1948.

Tennyson, Alfred. "Ulysses." *Tennyson: Poems and Plays.* London, New York, Toronto: Oxford University Press. 1973.

"The Wright Brothers." Dayton: Carillon Park. 1949.

This Fabulous Century, 1870-1900. New York: Time-Life Books. 1970.

Villard, Henry Serrano. *Contact: The Story of the Early Birds.* New York: Bonanza Books. 1968.

Walsh, John E. *One Day at Kitty Hawk.* New York: The Thomas Y. Crowell Company. 1975.

Washington Herald. Feb. 11, 1910. p. 1.

Whelan, Bernard L. "Wright Brothers Ran a School for 'Early Birds.' " *Dayton Journal Herald.* March 22, 1960. p. 5.

Wright, Milton. *Diary, 1855-1917.* Wright State University Archives.

Wright, Orville. "Last Will and Testament." Montgomery County, Ohio, Probate Court files, 1948.

Wilbur & Orville Wright A Bibliography. Arthur G. Renstrom, compiler. Washington: Library of Congress. 1968.

Wilbur & Orville Wright A Chronology. Arthur G. Renstrom, editor. Washington: Library of Congress. 1975.

"Wright Brothers Scrapbook, 1876—." Maintained by Dayton and Montgomery County Public Library.

Wright, Orville. "Diary of the First Flight." *Collier's,* Vol. 122. pp. 32-33. Dec. 1913.

————. "How We Made the First Flight." *Flying.* Vol. 2. pp. 10-12, 35-36. Dec. 1913.

————. "How We Invented the Aeroplane." Fred C. Kelly, editor. New York: David McKay. 1953.

————. "Scrapbooks." Washington: Library of Congress shelves.

Wright, Wilbur. "Experiments and Observations in Soaring Flight." *Journal of the Western Society of Engineers.* August 1903. Vol. 8. pp. 400-417.

————. "Our Mother." *West Side News.* July 8, 1889. p. 2.

————. "Last Will and Testament." Montgomery County, Ohio, Probate Court Files. 1912.

Wright. Wilbur and Orville. *Miracle at Kitty Hawk: The Letters of Wilbur and Orville Wright.* Fred C. Kelly, editor. New York: Farrar, Straus and Young. 1951.

————. "The Papers of Wilbur and Orville Wright." Washington: Library of Congress. Manuscript Division.

————. *The Papers of Wilbur and Orville Wright, Including the Chanute-Wright Letters and Other Papers of Octave Chanute.* Marvin W. McFarland, editor. New York, Toronto, London: McGraw-Hill Book Company, Inc. 1953. Two volumes.

Index

Z